Plants and the K-T Boundary

The Cretaceous Period of geologic time ended abruptly about 65 million years ago with global extinctions of life in the sea and on land – most probably caused by a catastrophic meteorite impact. Although much popular interest has focused on the fate of the dinosaurs at that time, the plants that existed in Cretaceous time also underwent extensive and permanent changes, and they reveal much more about the nature of this devastating event.

In *Plants and the K-T Boundary*, two of the world's leading experts in the fields of palynology and paleobotany integrate historical records and the latest research to provide a comprehensive account of the fate of land plants during this 'great extinction.' The book begins with chapters on how the geological time boundary between the Cretaceous and Paleogene periods (the K-T boundary) is recognized with varying degrees of resolution, and how fossil plants can be used to understand global events some 65 million years ago. Subsequent chapters present detailed evidence from case studies in over 100 localities around the world, including North America, China, Russia, and New Zealand. The book concludes with an evaluation of the various scenarios for the cause of the K-T boundary event and its effects on floras of the past and the present.

This book is written for researchers and students in paleontology, botany, geology, and Earth history, and will be of interest to everyone who has been following the course of the extinction debate and the K-T boundary paradigm shift.

DOUGLAS J. NICHOLS is a Research Associate with the Department of Earth Sciences at the Denver Museum of Nature & Science and a Scientist Emeritus with the United States Geological Survey (USGS). He received his Ph.D. in geology from The Pennsylvania State University before pursuing a career that has included university teaching, the oil industry, and 30 years of research with the USGS. Dr Nichols is a palynologist, with research interests in the fossil pollen and spores of Upper Cretaceous and Paleogene rocks, with emphasis on biostratigraphy, paleoecology, evolution, and extinction events. In 2005 he received the Meritorious Service Award from the US Department of the Interior for his research on the biostratigraphy of nonmarine rocks and the Cretaceous-Paleogene (K–T) boundary in western North America. Dr Nichols is the author or coauthor of more than 140 scientific papers and has served as editor of the journals *Palynology* and *Cretaceous Research*.

KIRK R. JOHNSON is Vice President of Research & Collections and Chief Curator at the Denver Museum of Nature & Science (DMNS). He joined the DMNS in 1991 after earning his doctorate in geology and paleobotany at Yale University. Dr Johnson's research focuses on Late Cretaceous and early Paleogene fossil plants and landscapes of the Rocky Mountain region and is best known for his research on fossil plants, which is widely accepted as some of the most convincing support for the theory that an asteroid impact caused the extinction of the dinosaurs. He has published many popular and scientific articles on topics ranging from fossil plants and modern rainforests to the ecology of whales and walruses, and coauthored the books *Prehistoric Journey: A History of Life on Earth* and *Cruisin' the Fossil Freeway*.

Plants and the K–T Boundary

Douglas J. Nichols[1]

AND

Kirk R. Johnson[2]

Denver Museum of Nature & Science
[1] Research Associate
[2] Chief Curator & Vice President for Collections and Research

CAMBRIDGE
UNIVERSITY PRESS

CAMBRIDGE UNIVERSITY PRESS
Cambridge, New York, Melbourne, Madrid, Cape Town,
Singapore, São Paulo, Delhi, Tokyo, Mexico City

Cambridge University Press
The Edinburgh Building, Cambridge CB2 8RU, UK

Published in the United States of America by Cambridge University Press, New York

www.cambridge.org
Information on this title: www.cambridge.org/9780521305631

First published 2008
First paperback edition 2011

A catalogue record for this publication is available from the British Library

ISBN 978-0-521-83575-6 Hardback
ISBN 978-0-521-30563-1 Paperback

Dedicated in loving memory to
Beatrice Olmstead Nichols
who never doubted her son would go far
but perhaps did not envision travels eons back in time
and
Katie Jo Johnson
who was always amused that the son of a Katie
would study the K–T boundary

Contents

PART III   INTERPRETATIONS 215

Preface

We both have been fascinated by the Cretaceous–Paleogene (K–T) boundary since the late 1970s. The Alvarez discovery galvanized our individual research efforts and we have worked together on this problem since we met in 1983. Our research has focused on western North America, so our data and interpretations are largely based on field work and laboratory analyses in this region. We have also studied terrestrial K–T boundary sections in Russia, China, and New Zealand and searched, unsuccessfully, for them in Mongolia, Patagonia, and India.

In preparation for writing this book, we made a comprehensive survey of the world's scientific literature through 2006 pertaining to plants and the K–T boundary. Our bibliographic database includes more than 500 references, but we have chosen to cite only those most relevant to understanding the effects of the terminal Cretaceous event on plants. We sought to interpret objectively the data available in those publications rather than simply to repeat the conclusions of the original authors. In many instances we agree with the original authors, but in some we do not. In the latter instances, we trust we have fairly presented their views and that we have given no reason for offense in our reinterpretation.

To present a major conclusion at the outset, we deduce that the changes in plant communities that took place at the K–T boundary are inextricably and causally linked to the impact of an extraterrestrial body on the Earth in the Caribbean region – the Chicxulub impact. The events that transpired in latest Cretaceous and earliest Paleogene time are much more complex than this statement suggests, however. We invite the reader to explore this fascinating subject with us.

We gratefully acknowledge the support and assistance of many colleagues over the last 25 years for their insights, inspiration, and collaborations in reference to the terrestrial K–T boundary. Specifically, we thank Walter Alvarez, David Archibald, Frank Asaro, Moses Attrep Jr., Richard Barclay, Edward Belt, Samuel

Bowring, Dennis Braman, William Cobban, Philip Currie, Robyn Burnham, Steven D'Hondt, Erling Dorf, Beth Ellis, David Fastovsky, Farley Fleming, Joseph Hartman, Leo Hickey, Jason Hicks, Brian Huber, Steven Manchester, Edward Murphy, John Obradovich, Carl Orth, Dean Pearson, Herman Pfefferkorn, Charles Pillmore, Robert Raynolds, Michele Reynolds, Eugene Shoemaker, Arthur Sweet, Louis Taylor, Alfred Traverse, Robert and Bernadine Tschudy, Gary Upchurch Jr., Vivi Vajda, Wesley Wehr, Peter Wilf, Scott Wing, and Jack Wolfe. Our field efforts outside of North America were supported or enlightened by Joan Esterle, Elizabeth Kennedy, Jane Newman, Ian Raine, Khishigjav Tsogtbataar, Mahito Watabe, Makoto Manabe, Masaki Matsukawa, Harufumi Nishida, Kaz Uemura, Rubén Cúneo, Alejandra Gandolfo, Rosendo Pascual, Pablo Puerta, Raminder Loyal, Ashok Sahni, Chen Pei-ji, Sun Chunlin, Sun Ge, Rahman Ashraf, Mikhail Akmetiev, Eugenia Bugdaeva, Lena Golovneva, Alexei Herman, Tatiana Kezina, Tatiana Kodrul, Valentin Krassilov, and Valentina Markevich. Beth Ellis prepared many of the maps and plates; Dennis Braman of the Royal Tyrrell Museum reviewed data on Canadian localities summarized in the Appendix; Ian Miller reviewed the final manuscript. We are grateful to the editorial staff of Cambridge University Press for their patience, support, and guidance.

DOUGLAS J. NICHOLS
KIRK R. JOHNSON

PART I   BACKGROUND

1

Introduction

1.1   The K–T controversy and the Alvarez challenge

A paper published in 1980 in the journal *Science* revolutionized the
science of geology. Coauthored by Nobel laureate in physics Luis Alvarez, his
geophysicist son Walter, and two colleagues, the paper presented data from
the esoteric field of neutron activation analysis. These data suggested that the
Earth had been struck by a large extraterrestrial object (an asteroid or possibly
a comet) some 65 million years ago, precisely at the moment in time that
marked the boundary between the Mesozoic and Cenozoic eras (Figure 1.1).
The time line, on a smaller scale also the boundary between the Cretaceous
and Paleogene periods, was widely known as the K–T boundary ("K" being the
internationally accepted abbreviation for Cretaceous and "T" being the corresponding abbreviation for either Tertiary or Paleogene, according to nomenclatural preference). The paper (Alvarez *et al.* 1980) also proposed that this
extraterrestrial impact had been responsible for one of the greatest episodes
of extinction in Earth history. The K–T extinctions, which eradicated 70% or
more of species on land and in the sea, ended the Mesozoic Era, the second of the
three great subdivisions of life recognized by paleontologists. The cause of the
K–T extinctions had long been argued in paleontology. The impact hypothesis
had now been put forward as the explanation.

The asteroid impact hypothesis, involving as it does a causative agent from
outside the Earth and also an instantaneous catastrophic event (an anathema in
geology in 1980), immediately became enormously controversial in paleontology and geology. To no small extent the controversy may have arisen also
because an explanation for the great extinction had been presented not by
paleontologists, who were inclined toward ownership of the phenomenon of

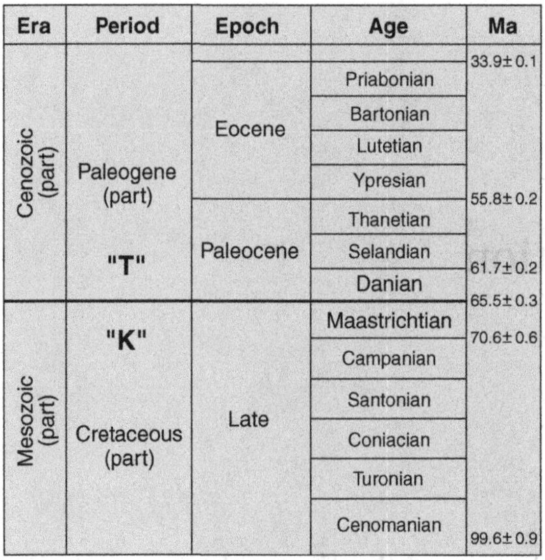

Era	Period	Epoch	Age	Ma
Cenozoic (part)	Paleogene (part)	Eocene	Priabonian	33.9± 0.1
			Bartonian	
			Lutetian	
			Ypresian	55.8± 0.2
	"T"	Paleocene	Thanetian	
			Selandian	61.7± 0.2
			Danian	65.5± 0.3
Mesozoic (part)	"K"	Late	Maastrichtian	70.6± 0.6
			Campanian	
	Cretaceous (part)		Santonian	
			Coniacian	
			Turonian	
			Cenomanian	99.6± 0.9

Figure 1.1 Part of the geologic time scale centered on the K–T boundary. Selected radiometric ages provide calibration (Ma = mega-annum, million years). Age data from Gradstein *et al.* (2004).

extinction, but by a physicist and his colleagues, outsiders as it were. Not all paleontologists took umbrage, but many did. The asteroid impact hypothesis soon became widely known in the popular press because the best known of all fossil creatures – the dinosaurs – top the long list of organisms whose geologic history ended at the K–T boundary (indeed, the Mesozoic Era is popularly known as the Age of Dinosaurs). In the best tradition of the natural sciences, however, the impact hypothesis quickly generated a wide variety of studies designed to investigate its possible validity, or in many instances, intending to disprove it. These studies, far too numerous to be reviewed here, were responses to what we call the Alvarez challenge: to prove or disprove the impact hypothesis of the K–T extinctions. Because the matter of extinction largely concerns the field of paleontology, paleontologists – we among them – were those primarily challenged. The Alvarez asteroid impact hypothesis posed a specific question: did an extraterrestrial impact cause extinction? The challenge to us was to determine whether the fossil record of plants in terrestrial rocks could answer this question.

Throughout the 1980s, the evidence for the hypothesized impact grew. Anomalous concentrations of the metallic element iridium (Ir) at the K–T boundary, the Alvarez team's primary evidence, were located at dozens of new K–T sites and cores in marine and nonmarine (terrestrial) rocks around the world. The discovery of impact-sourced shock-metamorphosed mineral grains at numerous K–T boundary sections overwhelmed the competing idea that volcanism in India was responsible for the iridium anomalies. The impact crater itself was eventually located and identified (Hildebrand *et al.* 1991). A full review of the fascinating story of the evolution of the impact hypothesis to the status of a scientific theory is beyond the scope of this book; the best accounts are Alvarez (1997) and Powell (1998), both eminently readable books in the history of science.

This book is our answer to the Alvarez challenge. Popular interest in dinosaurs notwithstanding, fossil plants yield the most information about the effects of the K–T extinction event on the land. Our goal is to summarize evidence from fossil plants that bears on the impact extinction theory.

1.2   The central role of plants as evidence

As succinctly stated by Hickey (1984), land plants form a central element in any comprehensive inquiry into possible causes of extinctions at the K–T boundary because they are a conspicuous and exposed part of the terrestrial biota. Plants are speciose and common on terrestrial landscapes. They are primary producers, composing the base of the food chain. The fate of terrestrial animals depends upon them, either directly or indirectly, as food sources and for shelter; thus, the collapse of plant communities would cause the collapse of entire ecosystems. Unlike animals, plants are fixed in position on the landscape and cannot escape sudden deleterious changes in the environment. They are directly linked to atmospheric chemistry, temperature, and humidity and hence they reflect climate and are exquisitely sensitive to changes in it. Add to these essential aspects the fact that plants tend to be commonly preserved as fossils. Fossil plants also have biostratigraphic utility and they can be used to locate the stratigraphic position of the K–T boundary with great accuracy and precision. Thus, fossil plants are available for study, and they are the very organisms that have enormous potential for revealing critical information about the nature and effects of the K–T extinctions.

Plant fossils are preserved most often in two forms: as relatively large parts of whole plants ranging from leaves to tree trunks (plant megafossils); and as microscopic bodies such as pollen grains and spores (plant microfossils). Plant megafossils are common in the geologic record and plant microfossils are nearly ubiquitous in unoxidized, fine-grained, sedimentary rocks. As is discussed at greater length in Sections 3.1 and 3.2, plant megafossils – principally leaves – yield much valuable data about their geologic age, depositional environment, source vegetation, paleoclimate (especially temperature and humidity), and their insect herbivores; and plant microfossils, by virtue of their near ubiquity,

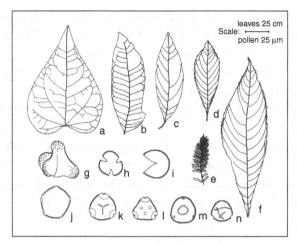

Figure 1.2 Sketches of typical fossil leaves and pollen from Maastrichtian, Paleocene, and lower Eocene strata in Montana and Wyoming, USA (from Nichols *et al.* 1988). a – *Paranymphaea crassifolia*, b – aff. *Averrhoites*, c – Lauraceae, d – *Pterocarya glabra*, e – *Metasequoia occidentalis*, f – "*Carya*" *antiquorum*, g – *Aquilapollenites quadrilobus*, h – *Tricolpites microreticulatus*, i – *Taxodiaceaepollenites hiatus*, j – *Polyatriopollenites vermontensis*, k – *Momipites leboensis*, l – *Momipites triorbicularis*, m – *Caryapollenites veripites*, n – *Platycarya platycaryoides*. Reprinted with permission.

Figure 1.3 Photograph of a representative outcrop of the kinds of rocks of varied fine- to coarse-grained lithology that can yield plant microfossils and megafossils. The exposure has enough lateral and vertical extent to yield several sections for analysis. This outcrop, at Clear Creek North in the Raton Basin (see Section 7.2), actually contains the K–T boundary (see arrow). Photo by C. L. Pillmore, US Geological Survey.

excel at stratigraphic age determination and correlation. Even though a given source vegetation produces both megafossils and microfossils, the resulting fossil records often contain strikingly different information. Where it is possible to employ both of these major kinds of plant fossils in investigations of the K–T boundary, each group supplements the other in revealing the nature of ancient plant communities and their fate as a consequence of the K–T boundary impact event. Figure 1.2 illustrates some plant megafossils (leaves) and microfossils (pollen) as generalized examples. Figure 1.3 shows an outcrop section that has yielded both plant megafossils and microfossils near and at the K–T boundary, bearing evidence of events that affected plants in latest Cretaceous and earliest Paleocene time.

1.3   Expectations of how plants should respond to a global catastrophe

In analyzing the differences between the evolutionary history of plants and animals, Traverse (1988a) asserted that major plant extinctions have not been synchronous with animal extinctions in the geologic past, and that changes in floras through time have been due to gradual replacement, not mass extinction. He cogently argued that this is because as a group, plants are resilient organisms able to survive extrinsic stresses much better than animals. Reasons for this include plants' ubiquity on landscapes, their indeterminant growth, their ability to sprout new shoots from rootstocks, and the long temporal persistence of seed banks. Traverse (1988a) concluded that major extinction events among plants and animals might not be attributable to the same causes. The extrinsic stresses that Traverse referred to were defined by DiMichele *et al.* (1987) as those caused by external agents that alter prevailing conditions locally to globally; they cited the K–T impact as an example. Although we agree with many of Traverse's observations and conclusions, we must disagree with him about the K–T extinctions, which we assert simultaneously affected both plants and animals on land as well as diverse marine organisms.

The first requirement in substantiating our claim is to consider how plants would have responded to an extrinsic stress of global proportions, a disaster such as the impact of a large extraterrestrial body on the Earth. Following the publication of Alvarez *et al.* (1980), many scenarios were developed that proposed various dire effects of a large impact on the terrestrial environment. We evaluate some of these ancillary effects in Section 11.5. That impact–extinction model is: impact followed immediately by shock waves and possibly by extensive wildfires; an enormous cloud of dust raised by the impact reached a low

Earth orbit and spread around the Earth, severely reducing solar radiation for a period of a few years; loss of sunlight causing supression of photosynthesis, which created an ecological catastrophe resulting in major extinction. This mechanism would appear to be adequate to explain the extinction of a vast number of those plant taxa (species or genera) that could not regenerate from seeds or rootstocks.

It is reasonable to assume that within a certain proximity to the impact site, possibly thousands of kilometers, forest vegetation would have been razed by shock waves emanating from the impact blast, and perhaps much of the fallen vegetation would have been set afire (Wolbach et al. 1990, Melosh et al. 1990). As the impact dust eventually settled and the sky cleared, a devastated landscape would have been revealed. It is likely that some plants survived in refugia, places protected from blast effects and forest fires. Those plants could begin to revegetate the landscape, but quite likely the first plants to appear and proliferate would have been ferns, which are able to grow quickly from spores or buried rhizomes. Ferns were "disaster species" in the sense of Harries et al. (1996) and Kauffman and Harries (1996), able quickly to colonize disturbed terrain. After the K–T impact event, they would take temporary advantage of the absence of seed plants and would dominate the landscape as the pioneer plant community. This is essentially the scenario envisioned by Tschudy et al. (1984) and DiMichele et al. (1987).

The geologic record of such an event or series of events could be expected to be unmistakable. Plant extinctions would be indicated by the disappearance from the fossil record of a significant number of taxa. Assuming not just mass kill but also mass extinction, megafossil floras on either side of the K–T boundary would be strikingly different in composition, as would the corresponding microfossil floras. The microfossils would be expected to exhibit their most profound changes coincident with the deposits representing impact debris. Pioneer communities of ferns would be recorded in the stratigraphic record by unusual abundances of fern spores just above the level of pollen extinction. Although not all plant taxa were driven to extinction, some might be expected to show sharp changes in abundance, either reductions or (like the ferns) increases. Plant communities on continents farther removed from the impact site might suffer less than those in closer proximity. Long-term effects on the Earth's vegetation would be expected to involve permanent changes in the composition of the surviving flora on a regional to global scale and reflect the nature and rate of ecosystem recovery. The record of the event might well be clearer in some locations than in others, although a truly global catastrophe could be expected to have some truly global effects. Our survey of the record of plant fossils across the K–T boundary is worldwide so that this issue can begin to be addressed.

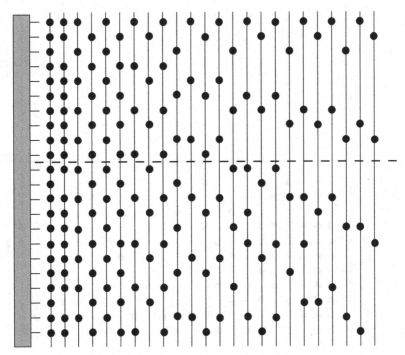

Figure 1.4 Diagram of a hypothetical, fossil-bearing stratigraphic section to illustrate the Signor–Lipps effect. Tick marks represent sample levels, vertical lines represent stratigraphic ranges of different fossil species having variable frequencies of occurrence in the section, and black dots represent presence of fossils (one or more specimens of a species) in samples. The dashed horizontal line represents the level at which abrupt and total extinction will be assumed to occur (compare Figure 1.5).

A most important consideration is not only how plants would respond to such a global catastrophe, but also how plant fossils would leave an interpretable record of the event. We speak here not of modes of preservation of plant fossils (those are briefly summarized in Sections 3.1 and 3.2), but of observations that have been made on the occurrence of fossils in general. It is well known that within any taxonomic group preserved in the fossil record, some taxa are abundant, others are less common, and some are rare. Stratigraphic occurrences of abundant, common, and rare taxa of any group of fossils within a stratigraphic interval create predictable patterns when plotted sample-by-sample. Figure 1.4 is a hypothetical example.

In Figure 1.4, the vertical column at the left represents a stratigraphic section from which fossils have been collected; tick marks show sample positions. Occurrences of individual taxa (species or genera) are shown by the black dots at the stratigraphic levels where they were found; each dot represents one or more specimens recovered at that level. The vertical lines represent the

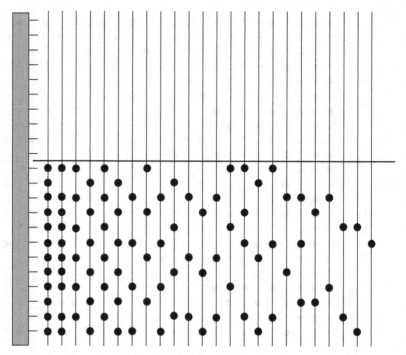

Figure 1.5 Diagram based on Figure 1.4 illustrating the Signor–Lipps effect. The solid horizontal line represents the extinction level after an extinction event. Compare Figure 1.4 and see text for further explanation.

stratigraphic ranges of the taxa in this section – note that all taxa range to the top or nearly to the top of the section. The taxa are arranged in order of abundance with some occurring at all or most levels (left side of diagram), others at fewer levels, and rare ones at very few levels (right side of diagram). In Figure 1.4, the horizontal dashed line within the stratigraphic sequence indicates the level at which a theoretical mass extinction of all the taxa will take place. In Figure 1.5, the pattern among the abundantly occurring taxa clearly shows that, at the level of the solid line, an abrupt and total extinction has occurred. However, the pattern among the less abundant and rare taxa fails to show clearly the level of extinction because the last occurrences of these taxa are well before the theoretical extinction level. Furthermore, the occurrences of the rare taxa, taken together, suggest that the mass extinction was not abrupt, but that it occurred gradually.

Signor and Lipps (1982) considered this phenomenon and formulated the concept that artificial range truncations (by which they meant the abrupt terminations of the stratigraphic ranges of taxa), especially among uncommon taxa, give the appearance of a gradual extinction even if the extinction is abrupt

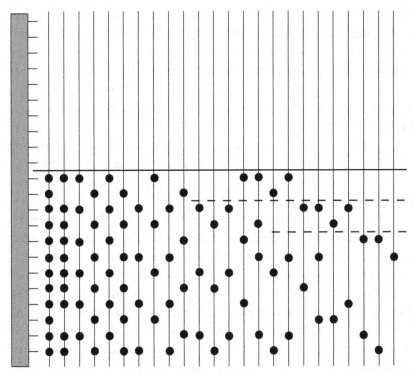

Figure 1.6 Diagram based on Figure 1.5 showing the actual extinction level (solid horizontal line) and apparent "stepwise" levels of extinction (dashed horizontal lines).

and catastrophic. The phenomenon has come to be known as "the Signor–Lipps effect." The Signor–Lipps effect makes it appear that even an abrupt and total extinction was to some extent gradual. Signor and Lipps noted that more extensive sampling could fill in some gaps in the stratigraphic ranges of some taxa, making a curve drawn on the last occurrences closer to a straight line that coincides with the extinction level; however, the curve would never flatten entirely.

Figure 1.6 is based on Figure 1.5, but in addition to the solid line marking the level of extinction as in Figure 1.5, dashed lines mark where it appears that substantial numbers of the taxa died out at levels or steps below the final extinction. However, this "stepwise" extinction pattern is merely an artifact of the varied relative abundances of the fossil taxa, as is the deceptive pattern of gradual extinction.

The Signor–Lipps effect, although a theoretical concept, was shown to be valid by Meldahl (1990). Meldahl conducted an experiment in which he used specimens of extant species of marine mollusks collected from Holocene beach sediments. The presence or absence of specimens at each sampling level was determined by the actual abundance of each species within the sampled interval. The surface of the beach represented a total catastrophic extinction because no specimens of any species were present above that level. When Meldahl plotted the data, he found that rare extant species seemed to disappear gradually below the simulated extinction level, the present-day beach surface. Meldahl's taxa were marine mollusks and his extinction level imaginary (because all his taxa are extant species), but the same Signor–Lipps effect can be anticipated for plant-fossil taxa at the actual extinction level of the K–T boundary. This is an important principal that we will return to later in discussing the plant fossil record at specific localities.

Our primary assertion is that, given an understanding of the preservation modes of plants and of the sampling effects involved in recovering extinctions from fossils, the plant fossil record, including both megafossils (especially leaves) and microfossils (especially pollen and spores), can yield invaluable information about Earth history. We believe that plant fossils are a largely unexploited key to understanding one of the most fascinating questions in geology, the nature of terminal Cretaceous extinctions. To comprehend the significance of plant fossils as they relate to the K–T boundary, it is necessary to appreciate how they are used to identify the boundary in conjunction with other geologic evidence, how these methods developed historically, and what paleobotany (the study of plant megafossils) and palynology (the study of plant microfossils) tell us about the vegetation of the Earth in Late Cretaceous and early Paleogene time. The chapters in Part I of this book address these essential matters, Part II presents case studies from those regions of the world for which data are available, and Part III covers broad-scale interpretations based on the data presented.

2

Resolution of the K–T boundary

We perceive three increasingly precise scales of temporal resolution of the K–T boundary. Stage-level resolution is on the order of millions of years; subchron-level resolution is on the order of tens to hundreds of thousands of years; and impactite-level resolution, while not directly measurable, is on the order of one to a few years. At the resolution of stage, the boundary between the Cretaceous and Paleogene periods is the boundary between the Maastrichtian Stage (5.1 Ma in duration) and the Danian Stage (3.8 Ma in duration). At the time of the Alvarez discovery, the resolution of most K–T boundary paleontological studies was at the stage level, at best. For most terrestrial K–T sections outside of North America, this level of temporal resolution is still prevalent today. At the subchron level of temporal resolution, the K–T boundary event occurred within the polarity subchron C29r, whose duration was between 570 and 833 thousand years, depending on which calibration is chosen. Resolution at the subchron level often brings recognition of paleontological events to the outcrop scale on the order of tens to hundreds of meters of section. Biostratigraphic zones can have similar durations to subchrons and thus may fall into this category of resolution. The third level of resolution is the impact layer, where physical and biological observations can be directly related to the K–T boundary impactite. At this level, events are resolved at the centimeter, or even millimeter, scale. A great deal of confusion and misunderstanding about the K–T boundary stems from the fact that different workers are mixing results and interpretations from different scales of temporal and stratigraphic resolution.

2.1   Stage-level resolution

Stages, the shortest formal chronostratigraphic units, are defined by paleontological content delimited at a stratotype. Stages are typically a few to

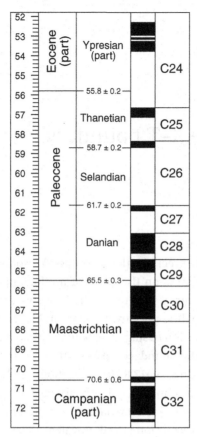

Figure 2.1 Subdivisions of the geologic time scale near the K–T boundary at the magnetic polarity subchron level. Normal polarity subchron intervals are shown in black and reversed polarity subchron intervals are shown in white. The K–T boundary is within C29r, the reversed part of Chron (C) 29. Dates in Ma. Data from Gradstein *et al.* (2004).

several million years in duration. The Maastrichtian Stage (the final stage of the Cretaceous, Figures 1.1 and 2.1), begins at 70.6 ± 0.6 Ma, based on correlation of its stratotype in southwestern France to the marine strontium isotope curve, and ends at the K–T boundary (Gradstein *et al.* 2004). All or part of six geomagnetic polarity subchrons occur in the Maastrichtian (part of C32n, C31r, C31n, C30r, C30n, and part of C29r). Because C30r is brief (∼10 thousand years), the 5.1 Ma of the Maastrichtian is essentially represented by a normal polarity interval bounded by two reversed polarity intervals (Figure 2.1). The Danian Stage (the first stage of the Paleocene, Figures 1.1 and 2.1) begins at 65.5 ± 0.3 Ma and ends at 61.7 ± 0.2 Ma (Gradstein *et al.* 2004). The Danian, when first defined by Desor (1847), was considered to be the last stage of the Cretaceous. This change has confused workers and the resulting terminology over the last century. Parts of six geomagnetic polarity subchrons occur in the Danian (part of C29r, C29n, C28r, C28n, C27r, and C27n).

Both the Maastrichtian and Danian are defined by their content of marine fossils and thus are terms that are best applied to marine rocks. In addition, the base of the Maastrichtian has been subject to many revisions and has recently ranged from 73 Ma (Harland *et al.* 1982) to 74 Ma (Harland *et al.* 1989) to its present definition of 70.6 Ma (Gradstein *et al.* 2004).

It has been, and continues to be, challenging to refer terrestrial strata to marine stages, and local terrestrial stages have been proposed to address this issue. In North America, a system of North American Land Mammal Ages was defined by the Wood Commission in 1941 to correlate the extensive, mammal-bearing, terrestrial sequences of the Cenozoic. Correlation of terrestrial stages with the time scale has improved with the use of intercalated marine and nonmarine sections and the rise of independent dating techniques such as magnetostratigraphy and geochronology. In the Western Interior of North America, stage-level resolution of terrestrial K–T strata was achieved by con-straining the terrestrial K–T boundary interval between Maastrichtian marine mollusks in the Fox Hills Formation and Danian foraminifera in the Cannonball Formation (Fox and Ross 1942). If only stage-level resolution is in place, the precision of statements concerning K–T boundary events is limited. Ultimately, stage-level resolution rests upon the recognition of age-diagnostic fossils and in terrestrial sections these are usually represented by vertebrates, plants, and freshwater mollusks. There are still many terrestrial sections of Late Cretaceous and Paleocene age where stage-level correlation to the marine record has not been achieved and the resolution of K–T boundary discussions carries error bars of several million years.

2.2   Subchron-level resolution

Magnetostratigraphy allows for marine–nonmarine correlation at higher resolution than at the stage level (Figure 2.1). Magnetic polarity sub-chrons in the Upper Cretaceous range from very short duration (C30r is esti-mated at 10 000 years) to very long (the Cretaceous Long Normal C34n lasted more than 40 million years), but the 11 subchrons of the Maastrichtian–Danian average about 800 thousand years (Cande and Kent 1995) and C29r, which spans the K–T boundary, is between 570 and 833 thousand years (Cande and Kent 1995, D'Hondt *et al.* 1996). Thus, the addition of magnetostratigraphy to terrestrial sections can tighten resolution to within a million years and thus bring correlation to the outcrop scale. In many terrestrial sections, sediment accumulation rate is on the order of tens to hundreds of meters per million years, and individual subchrons can be measured in a single stratigraphic sequence. In both nonmarine and marine sections, magnetostratigraphy relies upon biostratigraphy or geochronology to place normal and reversed paleomagnetic intervals within a chronostratigraphic framework. Examples of applications of magnetostratigraphy to K–T boundary studies in nonmarine rocks include Lerbekmo and Coulter (1984), Lerbekmo (1985), Erben *et al.* (1995), Hicks *et al.* (2002, 2003), and Barclay *et al.* (2003). Because the duration of polarity subchron C29r is estimated at 570 to 833 thousand years and the K–T boundary in marine rocks occurs in the upper half of the subchron, magnetostratigraphy can be used to locate a potential K–T boundary to within a half-million years.

Geochronologically, the age of the K–T boundary is estimated at 65.5 ± 0.3 Ma (Gradstein *et al.* 2004). This estimate is based on several $^{40}Ar/^{39}Ar$ ages derived from Haitian tektites and sanidine crystals from K–T boundary clay layers in Montana. Obradovich in Hicks *et al.* (2002) recalculated the K–T age at 65.51 ±0.1 Ma. Given that U-Pb ages are typically 0.5% older than $^{40}Ar/^{39}Ar$ ages from the same rock, U-Pb estimates of the age of the K–T will likely be around 66 Ma (note that the difference between the two systems is currently much greater than the precision within either system). As currently practiced these geochronologic techniques offer temporal resolution on the same level as or slightly more precise than the geomagnetic time scale. Recent developments suggest that geochronology will soon offer resolution on the order of less than 20 thousand years. If realized, this will offer a level of resolution intermediate between subchron and impactite. Resolution at the 20 000-year-scale is now available from cyclostratigraphy but is largely restricted to marine strata and thus is only indirectly relevant to our study.

2.3   Impactite-level resolution

The Alvarez discovery led to the ultimate temporal resolution currently possible in the rock record. Making the well-supported assumption that the K–T boundary recognized by Alvarez is an impactite allows global marine and nonmarine correlation at resolution of less than one year (Figure 2.2). This was a paradigm-altering breakthrough at a time when one-million-year resolution was not guaranteed by any other method. In the future, perhaps one of the most important uses of the K–T boundary will be as a test of high-resolution geochronology.

In the best-preserved K–T boundary sections (our discussions do not include K–T impactites preserved in marine strata), the extinction horizon and the anomalous concentration of iridium are found in association with a unique claystone layer of kaolinitic and smectitic composition. The concentration of

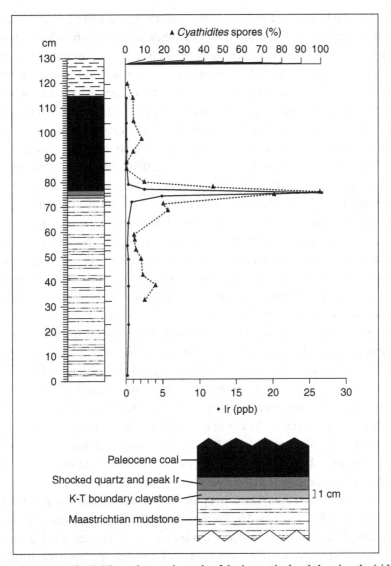

Figure 2.2 The K–T boundary at the scale of the impactite level showing the iridium anomaly and fern-spore spike at the Sussex locality in the Powder River Basin (see Section 7.4). Enlarged detail below shows microstratigraphy of the boundary interval including mudstone bearing Maastrichtian pollen overlain by impactite deposits beneath coal of earliest Paleocene age. Peak iridium anomaly and most shocked quartz grains are present in carbonaceous mudstone above the K–T boundary claystone. Data from Nichols *et al.* (1992a).

iridium at the K–T boundary, which is measured by neutron activation analysis (e.g., Alvarez *et al.* 1980; Orth *et al.* 1981, 1982), is commonly referred to as the "iridium anomaly" because its abundance can be hundreds of times greater than the normal background level of this element in sedimentary rocks. The greatest iridium anomaly so far detected at the K–T boundary in nonmarine rocks is 71 nanograms per gram or parts per billion (ppb) (Vajda *et al.* 2001); measurable background levels of iridium in nonmarine rocks can be as low as 0.010 ppb (Orth *et al.* 1987). The claystone layer is interpreted to have originated as impact ejecta that was diagenetically altered to clay (Hildebrand and Boynton 1988, Smit 1999). Good examples of the K–T boundary claystone layer, which is 1–2 cm in thickness at most localities, are described and illustrated in Section 7.2.

Within or closely associated with the boundary claystone layer at many localities is mineralogic evidence of its impact origin: shock-metamorphosed mineral grains ("shocked quartz") and spherules. Shocked mineral grains exhibit multiple sets of parallel microfractures (Bohor *et al.* 1984, Izett 1990). The forces necessary to create such features are produced only by meteorite impacts or nuclear explosions and are never found in volcanic rocks (Izett 1990). Spherules are small glassy microtektites that cooled from melted rock that was ejected into the atmosphere by the impactor and then rained back to Earth. They are now altered to kaolinite or other minerals and are well known in marine sediments at the K–T boundary. Alteration tends to obliterate spherulitic structure in the boundary claystone, but well-preserved spherules are reported in nonmarine rocks at some localities (e.g., Bohor *et al.* 1987a).

A significant indicator of the K–T boundary is an anomalous concentration of fern spores just above the level of the extinction of Cretaceous pollen. A true fern-spore abundance anomaly or fern-spore "spike" is not simply a concentration of diverse fern spores. The fern-spore spike was originally observed by Orth *et al.* (1981) and first described by Tschudy *et al.* (1984). It was codified by Fleming and Nichols (1990), who formally defined it as a palynological assemblage composed of from 70 to 100% fern spores of a single species occurring within an interval 0–15 cm above the K–T boundary. The origin, expression, and interpretation of this phenomenon are discussed at length in Section 5.3. It is mentioned here because its presence may be used to confirm identification of the K–T boundary.

Some additional features of the K–T boundary at the impactite level of resolution have been reported in nonmarine rocks, but they have not been used to identify the boundary in sections where it was not already known. Shifts in ratios of the stable isotopes of carbon of about two per mil have been detected (Schimmelmann and DeNiro 1984, Beerling *et al.* 2001, Arens and Jahren 2000, Arens and Jahren 2002, Gardner and Gilmour 2002), and microscopic diamond crystals (microdiamonds) presumably derived from the K–T impactor have been reported at a few localities (Carlisle and Braman 1991, Gilmour *et al.* 1992, Hough *et al.* 1995, Carlisle 1995).

By definition, the K–T boundary is represented by the Global Stratotype Section and Point (GSSP) established by the International Commission on Stratigraphy at El Kef in Tunisia, a site that occurs in marine rocks but also contains terrestrial palynomorphs. Terrestrial K–T boundary sections can only be validated if they can be shown to be coeval with the K–T boundary stratotype section. The burden of proving synchronicity is particularly heavy in the case of the K–T boundary, for which the putative cause is a bolide impact that occurred in a literal instant. In the best cases, those employing high-resolution geochronology or marine cyclostratigraphy, temporal resolution for the end-Cretaceous is about 20 thousand years. This represents one Milankovitch cycle in cyclostratigraphy or a one-sigma error of 0.04% in either the U-Pb or the $^{40}Ar/^{39}Ar$ decay series. It also is equivalent to the length of a typical Pleistocene interglacial interval. This is an inconvenient but critical fact: we do not have the direct temporal resolution to measure K–T boundary events on the time scales at which they occurred. Despite this limitation, we do have the ability to look at the relative timing of events in a microstratigraphic sense, and there are reasonable inferences that can be drawn from the nature of the stratigraphic record.

All K–T boundary studies are hampered by the resolution of time in the rocks being studied, but some are much more challenged than others. Studies that only achieve the stage or subchron level of resolution are common, and the results and interpretations must be taken in this light. One could make the argument that only the studies that achieve all three levels of resolution are directly germane to understanding the K–T boundary event, but because many areas around the world have yet to achieve triple-resolution status, we feel it is important to identify and discuss them for their future potential. The global distribution of published K–T boundary localities in terrestrial rocks is shown in Figure 2.3.

The study of the K–T boundary in terrestrial rocks is manifestly multidisciplinary. Therefore, there is tremendous potential for collaboration and understanding, but also the risk inherent in any collaborative chain: the conclusions are only as strong as the weakest link. It is for this reason that we take pains to define a multidisciplinary "tool kit" with which we will evaluate terrestrial K–T boundary sections and studies. Only by accounting for the vagaries of stratigraphy, sedimentology, taphonomy, geochronology, magnetostratigraphy, impact stratigraphy, paleobotany, and palynology can we expect to retrieve a reliable record of what happened on the planet 65.5 million years ago. The K–T boundary tool kit used to define the impactite-level resolution includes the following

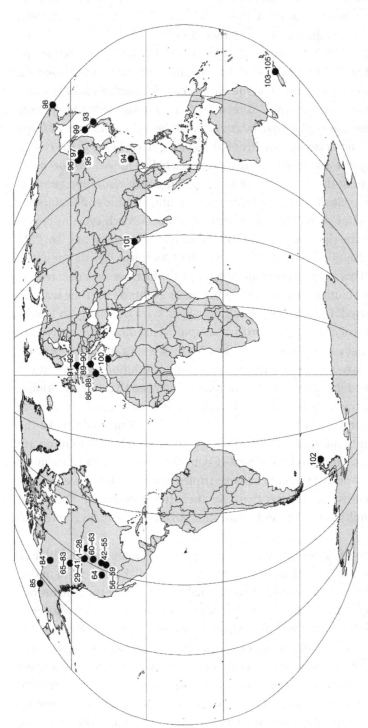

Figure 2.3 Map of the world showing the locations of the K–T boundary sections in nonmarine rocks listed in Table 2.1 and Appendix by locality numbers.

Table 2.1 *Published K-T boundary sections in terrestrial rocks*

Region, Appendix no., and Name	Palyno	Sub-dm	Ir ppb	Clayst	Sh qtz	Pmag	Isoage	K vert	T vert	K flora	T flora	F spike	TOTAL
Williston Basin, SW North Dakota													
(1) Pyramid Butte	1	3	3		3	1		1	1	1	1		15
(2) Sunset Butte	1					1		1	1	1	1		6
(3) River Section	1					1		1	1	1	1		6
(4) Bobcat Butte	1	3				1		1	1	1	1		9
(5) Mud Buttes	1	3	3	3	3	1		1	1	1	1	1	19
(6) Pretty Butte	1					1		1	1	1	1		6
(7) Cannonball Creek	1					1		1	1	1	1		6
(8) Big Boundary	1					1		1	1	1	1		6
(9) Pretty Butte North	1					1		1	1	1	1		6
(10) Terry's Fort Union Dinosaur	1	3				1		1	1	1	1		9
(11) Mikey's Delite	1	3				1		1	1	1	1		9
(12) Dean's High Dinosaur	1	3				1		1	1	1	1		9
(13) Scorpion Stung Kristian	1	3				1		1	1	1	1		9
(14) Vertical Doug	1					1		1	1	1	1		6
(15) Badland Draw	1	3				1		1	1	1	1		9
(16) New Facet Boundary	1	3				1		1	1	1	1		9
(17) Torosaurus Section	1	3				1		1	1	1	1		9
Williston Basin, central North Dakota													
(18) Huff	1					1							2
(19) Katus Site	1						1						2
(20) Brenner Site	1												1
(21) Cannonball Stage Stop Site	1												1
(22) Rattlesnake Butte Site	1												1
(23) Schaeffer Site	1												1
(24) Knispel Site	1												1

Table 2.1 (cont.)

Region, Appendix no., and Name	Palyno	Sub-dm	Ir ppb	Clayst	Sh qtz	Pmag	Isoage	K vert	T vert	K flora	T flora	F spike	TOTAL
(25) Miller Site	1					1							2
(26) Stumpf Site	1						1						2
(27) University of Mary Site	1												1
(28) Snyder Site	1												1
Williston Basin, eastern Montana													
(29) Glendive	1					1		1	1			1	5
(30) Brockton	1						1	1					2
Hell Creek area, Montana													
(31) Brownie Butte [1970]	1				3	1	1	1			1		8
(32) Brownie Butte [1984]	1	3	3	3	3	1	1	1			1	1	18
(33) Brownie Butte [1999]	1	3	3	3	3	1	1	1			1	1	18
(34) Herpijunk Promontory	1	3	3	3			1	1				1	13
(35) Herpijunk Northeast	1	3	3	3			1	1				1	13
(36) Billy Creek	1	3	3	3		1	1	1				1	14
(37) Rick's Place	1	3	3	3			1	1				1	13
(38) Iridium Hill	1	3	3	3			1	1				1	13
(39) Lerbekmo	1	3	3	3			1	1				1	13
(40) Seven Blackfoot	1	3	3	3			1	1				1	13
(41) Seven Blackfoot Creek	1			3				1				1	6
Raton Basin, New Mexico													
(42) York Canyon Core	1	3	3	3		1						1	12
(43) City of Raton	1	3	3	3	3								13
(44) Sugarite	1	3	3	3	3							1	14
(45) North Ponil	1	3	3	3	3								13
(46) Dawson North	1	3	3	3								1	11
(47) Crow Creek	1	3	3	3									10

Raton Basin, Colorado

No.	Locality													n
(48)	Starkville North	1	3	3	3	3							1	14
(49)	Starkville South	1	3	3	3	3							1	14
(50)	Clear Creek North	1	3	3	3	3								13
(51)	Clear Creek South	1	3	3	3	3								13
(52)	Madrid	1	3	3	3	3								13
(53)	Berwind Canyon	1	3	3	3	3							1	14
(54)	Long Canyon	1	3	3	3								1	11
(55)	Carmel		3	3	3	3								9

Denver Basin, Colorado

No.	Locality													n
(56)	South Table Mountain	1						1		1	1	1		6
(57)	Kiowa Core	1	3				1				1	1		10
(58)	Castle Pines Core	1					1							3
(59)	West Bijou Site	1	3	3	3	3	1				1	1	1	20

Powder River Basin, Wyoming

No.	Locality													n
(60)	Type Lance Area	1						1	1					3
(61)	Dogie Creek	1	3	3	3	3		1	1				1	16
(62)	Teapot Dome	1	3	3	3	3		1		1			1	16
(63)	Sussex	1	3	3	3	3			1				1	15

Wasatch Plateau, Utah

No.	Locality													n
(64)	North Horn Mountain							1	1	1				2

Western Canada

No.	Locality													n
(65)	Red Deer Valley	1	3	3	3	3	1	1	1					13
(66)	Coal Valley	1	3	3			1							7
(67)	Judy Creek 83-313A	1	3	3	3	3								8
(68)	Judy Creek 83-368A	1	3	3	3	3								8
(69)	Judy Creek 83-401A	1	3											5
(70)	Knudsen's Farm	1	3	3	3		1						1	15
(71)	Knudsen's Coulee	1	3	3	3		1						1	9

Table 2.1 (*cont.*)

Region, Appendix no., and Name	Palyno	Sub-dm	Ir ppb	Clayst	Sh qtz	Pmag	Isoage	K vert	T vert	K flora	T flora	F spike	TOTAL
(72) Hand Hills	1			3									4
(73) Castle River	1	3	3	3									10
(74) Ravenscrag Butte	1	3	3			1	1				1		10
(75) Frenchman Valley	1	3	3	3		1	1					1	13
(76) Morgan Creek	1	3	3	3	3	1						1	15
(77) Rock Creek East A	1	3	3	3								1	11
(78) Rock Creek West A	1	3	3	3								1	11
(79) Rock Creek West B	1	3		3								1	8
(80) Rock Creek West C	1			3									4
(81) Rock Creek West E	1	3	3	3								1	11
(82) Wood Mountain Creek	1	3	3	3								1	11
(83) CCDP Core 13–31–1–2 W3	1	3	3	3								1	11
(84) Police Island	1	3	3										7
Alaska													
(85) Ugnu SWPT-1 Core	1	3											4
Localities outside of North America													
Spain													
(86) Coll de Nargo	1												1
(87) Fontllonga	1					1		1	1		1		5
(88) Campo	1					1		1					3
France													
(89) Rousset	1							1					2
(90) Albas	1												1
Netherlands													
(91) Geulhemmerberg	1	3											4
(92) Curfs Quarry	1												1

#	Locality	Palyno	Sub-dm	Sh qtz	Clayst	Ir ppb	Pmag	Isoage	K vert	K flora	T vert	T flora	F spike	TOTAL
Japan														
(93)	Kawaruppu	1			3								1	5
China														
(94)	Nanxiong	1					1			1	1	1		5
(95)	Baishantou	1								1		1		3
Russia														
(96)	Blagoveshchensk	1								1				2
(97)	Kundur	1								1	1	1		4
(98)	Beringovskoe	1								1	1	1		4
(99)	Sinegorsk	1								1				2
Africa														
(100)	El Kef	1			3	3								7
India														
(101)	Anjar	1			3		1							5
Antarctica														
(102)	Seymour Island	1			3									4
New Zealand														
(103)	Moody Creek Mine	1			3	3		1				1	1	10
(104)	Compressor Creek	1			3			1				1	1	7
(105)	Mid-Waipara River	1			3	3						1	1	9

Localities are grouped geographically and numbered sequentially. Details for each are given in Appendix 1. Key to abbreviations: **Palyno**, palynofloral record across the putative boundary; **Sub-dm**, sample intervals of one decimeter or less; **Ir ppb**, iridium anomaly present; **Clayst**, boundary claystone layer present; **Sh qtz**, shocked minerals present; **Pmag**, magnetostratigraphic data available; **Isoage**, isotopic age data available; **K vert**, Maastrichtian vertebrate fauna below; **T vert**, Paleocene vertebrate fauna above; **K flora**, Maastrichtian megaflora below; **T flora**, Paleocene megaflora above; **F spike**, fern-spore abundance anomaly present. Features **Sub-dm, Ir ppb, Clayst,** and **Sh qtz** are triple-weighted because of their critical importance in identifying the K–T boundary. A **TOTAL** score of less than 10 indicates that the section probably is invalid for analysis of the record of plants at the K–T boundary. Note: The sections at El Kef (stratotype for the K–T boundary), Campo, Kawaruppu, Seymour Island, and Mid-Waipara River are marine sections for which data on terrestrial palynomorphs are available.

elements: (1) physical evidence of an impact based on a distinctive sedimentary layer, typically less than 1.5 cm thick, containing some combination of an anomalous concentration of iridium, shock-metamorphosed minerals, and spherules or tektites; (2) occurrence of the physical evidence within a zone of reversed magnetic polarity correlating to polarity subchron C29r; and (3) a larger biostratigraphic context both in terms of associated terrestrial biota and in relation to a time-correlative marine biota.

We have tabulated published terrestrial, plant-bearing, K–T boundary sections and have noted which attributes characterize each section (Table 2.1 and Appendix). By assigning scores for the presence of specific attributes at each section, we are able to assess the relative quality of the sections on a scale from 1 (least complete) to 20 (most complete). By this assessment, we recognize 100 terrestrial K–T boundary sections and 5 more in marine environments for which data on plant fossils are available. The majority (85) are located in North America (Figure 2.3).

Although scoring terrestrial K–T boundary sections may seem somewhat arbitrary, it provides a quick means of measuring the distribution and quality of terrestrial K–T boundary sections around the globe. Because of their critical importance in pinpointing the K–T boundary, we have given triple-weighting to certain attributes that are linked with high-resolution study or that define the impactite layer itself. These include subdecimeter sample resolution, the presence of a boundary claystone layer, existence of an iridium anomaly, and presence of shock-metamorphosed minerals.

The K–T sections in Table 2.1 are numbered sequentially, and these numbers will apply to the sections throughout the remainder of this book and will be used to identify localities on maps and figures.

3

Using fossil plants to study
the K–T boundary

3.1 The study of palynoflora

Palynofloras are essential to studies of the K–T boundary in nonmarine rocks. Numerous aspects of the nature of palynofloras must be considered when using them in such studies. These include: taxonomy (pollen or spore species vs. botanical species), preservation and preparation, sample size, facies effects, reworking or redeposition, stratigraphic resolution, and geographic coverage. These topics are covered in depth in textbooks on palynology (e.g., Traverse 1988b, Jansonius and McGregor 1996, Jones and Rowe 1999), but to be complete in our coverage of the use of palynofloras to study the K–T boundary, each of these topics is briefly discussed in this section.

Most genera of angiosperms (flowering plants) produce pollen with distinguishing morphologic features unique to those genera (number of apertures, exine structure and sculpture, size, etc.), which are the basis of palynological taxonomy. Most species within individual genera share the same characteristic features. Hence, whereas individual living and fossil genera can be differentiated, in most cases individual species cannot. Therefore, as a general rule, fossil pollen species can be thought of as representing botanical genera in the fossil record, and fossil pollen genera can be thought of as representing botanical families. This generality complicates the comparison between palynofloral and megafloral systematics. Comprehensive descriptions of the pollen morphology of living genera and species are in Edrtman (1965, 1966), and discussions and examples of the relationships of fossil pollen taxa to botanical taxa are in Muller (1970) and Traverse (1988b).

Palynomorphs (in nonmarine rocks: pollen of angiosperms and gymnosperms, spores of ferns and other cryptogams, and fungal spores) are organically preserved in fine-grained sedimentary rocks in vast numbers, especially in carbonaceous facies. They are acid-resistant and can be prepared for micro-scopic study by a variety of procedures usually involving disaggregation and chemical digestion of the sedimentary matrix. Details of what constitutes stan-dard palynological preparation techniques are described in Doher (1980) and Traverse (1988b). In palynologically fossiliferous rocks, usually only a few to several grams of sample will yield thousands of specimens. Palynomorphs are not abundantly or well preserved in sandstone or coarser clastic rocks or in any rocks that have been subjected to oxidative weathering. The deleterious effects of surface weathering must be kept in mind in sampling outcrops; drill cores usually obviate this potential difficulty by retrieving samples from well below the zone of surface weathering.

Palynomorphs are not uniformly distributed, even in fine-grained sedimentary facies. Lacustrine and mire deposits tend to preserve greater numbers of pollen and spores produced by the local flora of the depositional environment than do fluvial deposits. However, by virtue of the vast numbers of pollen and spores produced by local floras and their distribution to sites of deposition by wind and water, the regional as well as local vegetation usually is represented in the fossil record (Campbell 1999). Thus, palynomorphs are useful not only for paleoecolo-gical studies but for age determination and correlation. The relative abundance of various palynomorph taxa can vary in response to paleoenvironment and sedi-mentary facies. Therefore, in palynostratigraphic analysis, the presence or absence of a taxon most often is of greater significance than is its abundance.

Because of their small size and resistance to degradation, palynomorphs can be subject to reworking (emplacement in younger strata via erosion and redeposition). Specimens eroded from older deposits can be deposited in slightly younger or even much younger deposits, with or without contempora-neous microfossils. By virtue of their durability and the fact that they may be stratigraphically recycled enclosed in tiny chips of mudstone, reworked paly-nomorphs often show little evidence of being out of place, except perhaps from context. Reworking is an important issue that affects interpretation of micro-stratigraphic studies of the K–T boundary. Just as the Signor–Lipps effect can make an abrupt extinction appear gradual by smearing the record down-section, reworking can distort the record by smearing it up-section.

Because palynomorphs occur abundantly in suitable lithologies and can be recovered in significant numbers from small samples, stratigraphic resolution is possible at the centimeter and even millimeter scale. The stratigraphic resolu-tion possible using palynofloral analysis ultimately depends upon the precision of sampling, the quality of preservation of the fossils, and the thoroughness of taxonomic research that can be applied.

3.2   The study of megaflora

Megaflora includes identifiable macroscopic plant particles; the most commonly studied are leaves, fruits, seeds, and wood. To date, most work on the K–T boundary has focused on fossil leaf floras because leaves are the most commonly preserved macroscopic plant organs in this age rock (although European workers have focused more on fruits and seeds and Indian workers have concentrated on wood). Numerous aspects of fossil leaves must be considered when using them to study the K–T boundary. These include the origin of leaf litter and its relationship to its source vegetation; taphonomic and facies effects associated with transport, degradation, and burial; modes of leaf and cuticle preservation; sample size and sampling; stratigraphic and geographic spacing of sampled assemblages; taxonomic issues related to morphotaxa, leaf architecture, morphotyping, and the systematic utility of leaves; determination of paleoclimate from leaf physiognomy; and the utility of leaves for revealing plant–insect interactions.

Leaves are shed from trees in vast numbers. A typical forest tree will generate tens to hundreds of thousands of leaves per year resulting in millions of leaves per hectare (several hundred leaves per square meter) of forest floor for typical forests (Burnham and Wing 1989). Surprisingly, broadleaf evergreen forests tend to generate more leaf litter than deciduous forests, a fact attributable to the longer growing season and continuous leaf drop. Some herbaceous plants retain most of their leaves rather than dropping them, but all forests produce abundant litter. Forests growing on floodplains and other depositional environments produce litter that has high potential for burial either *in situ* from overbank flooding or by being entrained in channels.

Leaves and leaf litter degrade rapidly, and unless buried, will be fragmented or decayed to humus and recycled within months. Since sediment deposition in terrestrial ecosystems is typically episodic, the formation of leaf beds is more a function of rapid anoxic burial than the availability of leaves. Shallowly buried leaf beds are often destroyed by subsequent pedogenesis. For these reasons, leaf beds represent the least time-averaged of almost any type of fossil deposition, typically less than one year (Johnson 1993). Due to their abundance at the source, leaves generate rich fossil deposits and are suitable for the collection of large sample sizes.

Leaves do not fall far from their source tree and a majority of each year's litter is initially deposited within a distance from the trunk equal to its height. Comparison of the relative number of leaves from a single species in a forest and the basal area of trees of that same species has shown a positive correlation, so it is possible to recover the composition and relative abundance of forest trees from the leaf litter (Burnham and Wing 1989). Herbaceous plants are preserved only in exceptional cases such as volcanic ash falls or slurries (e.g., Wing *et al.* 1993). In many of these cases, whole plants are preserved.

Secondary transportation by wind or water can move leaves great distances and leaves falling into large rivers may travel hundreds of kilometers before downstream deposition. If a leaf assemblage can be shown to be *in situ* with careful taphonomic analysis, then the resulting flora, if properly sampled, can reflect the relative abundance of the source forest. Conversely, transported floras represent a broader but less precise sample of local vegetation.

Sedimentological analysis of fossil-bearing strata can identify original depositional settings and, used in concert with censuses of plant macrofossils, can provide the basis for mapping coeval vegetation types across ancient landscape surfaces (Hickey 1980). Because sedimentological facies evolve and migrate with basin evolution, it is critical to be aware of floral changes that are due solely to facies change.

Quality and type of preservation vary greatly from basin to basin and are linked closely to sediment grain size, diagenetic history, and basin chemistry. Some continental basins have been so oxidized that fossil leaves are effectively absent or, if present, are represented solely by faint impressions rather than compressions. Organic-rich sediments by definition preserve abundant plant fragments but the quality of preservation is linked to taphonomic history and degree of lithification. In the best cases, whole leaves are preserved with their waxy cuticle intact. In these cases, the fossil leaf provides dual information in the form of cuticular microstructure and leaf venation. In extremely rare cases, leaves are still attached to petioles and branches and these may be attached to fruits, flowers, cones, or seeds, providing an additional level of botanical information. In most cases, leaf venation is preserved in the absence of cuticle. A certain level of venation preservation quality is necessary for a fossil leaf to be identifiable. Angiosperm leaves have as many as seven orders of venation and at least four are desirable for reliable identification.

The identification of plant macrofossils is often an organ-specific exercise because plant parts are most often fossilized as detached organs. For this reason, leaves, fruits, seeds, and wood are often referred to morphotaxa, which are recognized in the International Code of Botanical Nomenclature (Greuter *et al.* 2000). To date, most Upper Cretaceous and Paleogene megafloral paleobotany has focused on fossil leaves and the utility of this system is increasing with developments in leaf architecture, which allow delineation of leaf taxa into morphotypes analogous to botanical species.

Leaves offer benefits in the form of climate proxy data from leaf physiognomy, which allow estimation of both mean annual temperature and precipitation (Wolfe 1993, Wilf 1997, Wilf *et al.* 1998), and plant–insect interaction data from insect-mediated leaf damage (Labandeira *et al.* 2002).

Based on censuses of leaf litter from extant forests, it is clear that more than 300 leaf fossils from a single quarry are desirable. This number allows for a reliable rarefied richness of the floras to be calculated using a rarefaction analysis of the relative abundance of taxa in the flora. Making a collection of this size usually represents a significant excavation and the removal of up to a cubic meter of rock. This is usually accomplished by the bench quarry technique where a leaf-bearing horizon is discovered and overburden is removed creating a benched surface on top of the productive horizon. A suite of prepared and censused fossils from a single bench quarry represents a sample that is tightly constrained in time and space, as it represents a group of fossils from a single deposit in a specific depositional environment. These collections, at their highest levels of precision, may sample as little as a few centimeters of strata and essentially represent less than a year's accumulation. These are arguably the most temporally precise of all fossil assemblages, greater perhaps even than pollen and spores, because leaves, as large and complex compressions, are not subject to stratigraphic reworking.

Because plant parts can be transported various distances, multiple samples from different facies at the same stratigraphic level can provide resolution of the original spatial heterogeneity of the source vegetation. In rare cases, leaf litter is even preserved in association with *in situ* trunks or stems and relatively specific vegetation patterns can be mapped (Ellis *et al.* 2003).

3.3   Relationship between palynoflora and megaflora

Even though leaves and palynomorphs are derived from the same source vegetation, their taxonomic resolution and taphonomic pathways are significantly different, such that they present different views of the same landscape. The strengths of palynology are limitless specimen numbers in the appropriate matrix, their ability to be retrieved from tiny rock chips such as well-cuttings and core samples, their ability to be sampled at very fine intervals, and their ability to travel into and be preserved in marine sediments. The weaknesses of palynology include a propensity of specimens to be reworked into younger strata by geologic processes, poor preservation in coarse or oxidized sediments, and relatively coarse taxonomic resolution. Fossil leaves provide high-resolution taxonomic data, climate proxies from physiognomy, and plant–insect interaction data from leaf damage. In addition, they are unlikely to be reworked into younger strata in a manner that is not immediately recognizable. The disadvantages of leaf records as opposed to palynological records are the relative difficulty of collecting leaves in great numbers and that stratigraphic resolution is rarely finer that a decimeter (usually a meter or more). However, leaves seem to travel less than palynomorphs and thus seem to map more tightly onto specific depositional environments. However, the majority of pollen grains also fall near their source plant (Kershaw and Strickland 1990), so it is possible to reconstruct local assemblages from both pollen and leaves.

In combination, leaves and palynomorphs have the potential to re-create excellent data concerning the occurrence of ancient plant taxa. Despite this, one of the great unresolved challenges is the correlation of pollen and leaf taxa. There are four main reasons for this discordance: (1) differential preservation of specific pollen or leaf taxa, (2) differential rates of leaf production by different plant habits (e.g., large trees vs. small herbs), (3) differential production of pollen and spores (e.g., wind vs. insect dispersal), and (4) differential taxonomic resolution of leaves and palynomorphs. It is well known that certain plant families have pollen grains with differential preservation potential and the same can be said of leaves. The family Lauraceae generally has tough and easily preserved leaves but easily degraded pollen. In contrast, herbaceous plants rarely shed leaves but often produce well-preserved pollen. In addition, certain plant habits (trees, vines) are more likely to shed greater numbers of leaves than are others (herbs and small shrubs), thus biasing the leaf fossil record toward arboreal forms in most cases. Pollen is abundantly produced by wind-dispersed plants but much less prolifically by insect-pollinated plants. These complex biases create a situation where the pollen and leaves in a particular stratum may be sampling different plant species and plant habits. For the Upper Cretaceous and Paleogene, this pattern is made more complicated by the fact that many of the taxa represent undescribed extinct families and genera and many of the leaves and pollen are not attributable to extant taxa or to each other. For all of these reasons, palynologists and paleobotanists have difficulty comparing their data and reaching conclusions based on integration of their data sets.

An example germane to this book is the calculation of extinction percentages based on leaves versus those based on palynomorphs. The closer sample spacing and large number of specimens per slide would seem to indicate that palynology would give a more accurate estimate of the severity of an extinction or extirpation. This appears not to be the case, however, because leaves have much higher taxonomic resolution. In thoroughly sampled sections, megafloral diversity exceeds palynofloral diversity despite a much smaller actual specimen count.

A corollary of this observation is that a palynological estimate of extinction severity will be less than the megafloral estimate.

Very few K–T boundary sections worldwide have benefited from joint mega- and microfloral investigation. Most of the case studies and sites discussed in this book consist solely of palynological data.

4

Brief history of K–T boundary paleobotany and palynology

4.1 First attempts in Europe and North America

The understanding of the nature of floral change across the K–T boundary has been delayed both by the nature of the plant fossil record and by poor temporal resolution in Cretaceous nonmarine rocks. The rapid rise of angiosperms in the Early Cretaceous and the rarity of thick, continuously fossiliferous, Cretaceous sections further complicated this situation. Angiosperms dominate modern vegetation and account for more than 80% of living species, but they appeared abruptly in the fossil record over a span of 25 million years in the Early Cretaceous. Early Cretaceous angiosperm leaves are superficially similar to living ones, and this similarity gave rise to the misconception that extant angiosperm genera first appeared in the Early Cretaceous. This sudden appearance, known as Darwin's abominable mystery, set the stage for several misconceptions and stratigraphic problems. Moreover, paleobotany has traditionally been summarized at the stage level and no great stock has been placed in obtaining the precise age of fossil floras. Before the Alvarez challenge, it was considered sufficient to state that a flora was Cenomanian or Campanian, or perhaps late Cenomanian or early Campanian. These stages are 6.1 and 12.9 million years in duration, respectively, and clearly represent too long of a time bin to be relevant to resolving change over short periods of time. Despite the fact that stages are too long to record rapid events, the nature of the terrestrial fossil record is such that even placing floras into a stage can be challenging when there is no direct correlation to marine strata. This situation has largely been resolved for Upper Cretaceous terrestrial rocks in the Rocky Mountain region of North America, but in other areas of the world paleobotany is still operating at the stage level of resolution and there are many regions where even stage-level resolution has not been achieved.

In the early days of paleobotany in Europe, this situation was complicated by the predominance of marine strata and the resulting discontinuity of the terrestrial Cretaceous rock record and the paucity of Upper Cretaceous and Paleogene plant-bearing beds. Whereas the marine invertebrate record of Europe showed a clear break at the K–T boundary with the distinct disappearance of ammonites and inoceramid bivalves, there were neither the terrestrial facies nor the fossils to track the history of plants over the same time period. The plant-bearing, Lower Cretaceous Wealden Group in the south of England contains no angiosperms. The most extensive Cretaceous angiosperm-bearing units in Europe, those in the Czech Republic, are Cenomanian, or roughly 30 million years older than the K–T boundary, but they contain abundant angiosperms and thus superficially appeared to have a modern aspect. The paleobotanist Oswald Heer made a practice of calling most angiosperm-dominated floras "Miocene" in a direct reference to their modern appearance. Similarly, Paleocene floras are also rare in Europe with only isolated sites near Gelendin in Belgium, Sezanne east of Paris, and the intertrappean deposits of the Isle of Mull in Scotland. Thus, the European record is markedly discontinuous and spotty on one hand and the early observers had not tuned their vision to distinguish between Cretaceous and Paleocene floras on the other. In fact, the Paleocene as an epoch, though defined in 1874, was not widely accepted until the 1950s (Mangin 1957).

In general, terrestrial sections can be relatively expanded compared with marine sections, yet individual exposures are rarely extremely thick, so the amount of time represented by any section is often relatively short. In the Rocky Mountain region, typical terrestrial sediment accumulation rates are about 100 m/my and exposures are typically tens to hundreds of meters thick. Thus, it is rare to find consecutive stages in superposition and fossil sites were usually discovered and studied in isolation from each other. For this reason, it was difficult to understand the temporal succession of fossil floras in advance of regional syntheses.

Such an opportunity was missed by Oswald Heer, who described dozens of floras that were discovered by Arctic explorers on Svalbard, Iceland, Greenland, and Alaska. The lack of vegetation cover over some of these sections presents more outcrop continuity than is found in the more vegetated lower latitudes. In his eight-volume *Flora Fossilis Arctica*, Heer described hundreds of leaf taxa from Cretaceous to Miocene strata. Unfortunately, very little stratigraphic information was retrieved and a golden opportunity to understand the succession of Arctic floras was delayed for more than a century. Some of Heer's collections from the Nûgssuaq Peninsula near Disko Island on west Greenland actually span the K–T boundary (Heer 1882, 1883; Koch 1964).

A clear understanding of the distribution of angiosperm-dominated floras relative to the K–T boundary did not begin until the exploration of the large sections of North America. Early work on the Albian–Cenomanian sequence exposed along the Atlantic coastal plain from New York to Virginia showed a sequence of floras that were admixed with bennettites, ferns, and conifers (Hickey and Doyle 1977). These rocks had similarities to the Wealden flora but they also contained angiosperms. A similar succession of rocks in the Black Hills of South Dakota included the transition from gymnosperm-pteridophyte floras in the Lakota Formation to ones dominated by angiosperms in the Dakota Sandstone (Ward 1899). Whereas exposures in the Atlantic coastal plain were heavily covered with vegetation and difficult to map, the sections in South Dakota were part of an extensive and easily mapped sequence that was continuous up through the Paleocene portion of the section. Lester Ward, Leo Lesquereux, and John Strong Newberry were quick to see that the angiosperm floras of the Dakota Sandstone were overlain by thousands of feet of ammonite- and marine-reptile-bearing Cretaceous marine rock and hence they deduced that the angiosperm radiation had occurred before the end of the Cretaceous.

4.2   The Laramie Problem in the American West

The sedimentary sequence of the Western Interior of North America was first mapped by the Hayden Surveys of the 1860s and 1870s as they worked their way up the Missouri River through South Dakota and into North Dakota and Montana. In this major transect, they recognized the broad outlines of the stratigraphy of the Western Interior seaway of the Cretaceous: a basal transgressive Dakota Sandstone overlain by a thick and fossiliferous sequence of marine shale and limestone (Pierre Shale and equivalents), overlain by a regressive Fox Hills Sandstone, overlain by a thick, lignite-bearing unit. The so-called "Great Lignite" contains units that we now know to be both Cretaceous and Paleogene in age, but at the time that was not clear. Superficially similar coal-bearing deposits were soon discovered in several stratigraphic units in additional areas in Wyoming, Colorado, Utah, and New Mexico. Eventually the Great Lignite came to be known as the Laramie Formation. The problem was that these coal-bearing units included rocks now known to be Late Cretaceous, Paleocene, and Eocene in age.

Discoveries of dinosaur fossils in the Great Lignite led to the realization that the K–T boundary was above the top of the Fox Hills Sandstone, but this realization took nearly 50 years to be resolved. The uncertainty over the age of the plant megafossils and the rocks that contained them continued until the middle of the twentieth century because of confused stratigraphy and flawed paleobotanical taxonomy. The stratigraphic confusion of the "Laramie

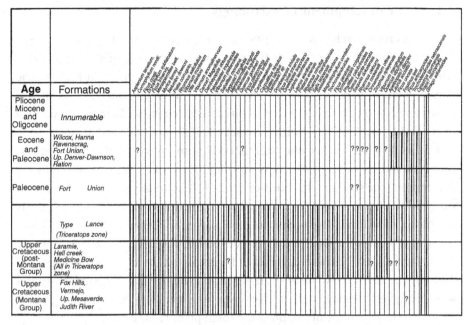

Figure 4.1 An early revelation of the magnitude of paleobotanical change across the K-T boundary (from Dorf 1940). Reprinted by permission.

Problem" and the inability to resolve angiosperm leaf taxa would diminish the utility of Cretaceous and Paleogene paleobotany until the late 1970s.

In 1922, paleobotanist F. H. Knowlton published a description of the mega-fossil flora of the Laramie Formation of the Denver Basin, Colorado. He had delayed publication for a number of years because of the confusion over the Laramie Problem and he included in his paleobotanical treatise an extensive and detailed review of the problem up until that time. In retrospect, the Laramie Problem can be thought of as a dispute between paleobotanists and vertebrate paleontologists about the correct stratigraphic position of the K-T boundary in western North America. Final resolution came from the careful analysis of both the stratigraphy and the megafossil floras by Dorf (1940, 1942a, 1942b) in the type area of the Lance Formation (uppermost Cretaceous) in the Powder River Basin, and by Brown (1943) in the Denver Basin. Dorf (1940) recognized a clear pattern of megaflora extinction across the K-T boundary but his analysis was at the formation level and he never clearly delineated an extinction event based on direct superposition of floras. Nonetheless, his 1940 diagram graphically shows for the first time the magnitude of the K-T megaflora extinction (Figure 4.1). This work was long ignored. Lingering problems in paleobotanical taxonomy were not clarified until still later with the advent of additional fossil leaf studies.

4.3   The development of palynology

As reviewed by Traverse (1988b), stratigraphic palynology had its beginning as a field of study (originally called pollen analysis) in the early 1900s
with the microscopic investigation of Pleistocene peat deposits in Europe. The
development of palynology in pre-Pleistocene stratigraphy began in the 1940s,
but its application to studies of Cretaceous and Paleogene rocks in North America
did not take place until later with the research of Rouse (1957) and Stanley (1965).
By the late 1960s it had become clear that palynomorphs, which occurred in vast
numbers and in seemingly endless morphologic variety, were useful tools for
determining the stratigraphic position of the K–T boundary (although the significance of that horizon with regard to a mass extinction event affecting plants was
unknown at the time). Two important papers were published in the early 1970s in
a *Geological Society of America Special Paper* devoted to the K–T boundary (Leffingwell
1970, Tschudy 1970); these papers will be discussed in detail later (Chapter 7).
Thus, by the 1970s the value of stratigraphic palynology in K–T boundary studies
was firmly established, and it was becoming evident that palynology also
increased the capability of sampling the fossil plant record.

Palynology helped to pinpoint specifically the stratigraphic position of the
K–T boundary in the early 1980s (Orth *et al.* 1981), which led to significant new
data on the effects of the K–T boundary event on plants (see Sections 6.3 and 7.2).
Palynology was used in conjunction with megafossil paleobotany at the K–T
boundary for the first time in the 1980s (Hickey 1981, Wolfe and Upchurch
1986, Johnson *et al.* 1989). Palynology is now the primary tool for locating the
K–T boundary in nonmarine rocks.

4.4   Leaf architecture and the systematic utility of leaves

American paleobotany had grown from European roots and its practi-
tioners followed the nomenclatural lead of Heer, Saporta, von Ettinghausen,
and Brongniart. As a result, the basionyms of many of the species described by
Lesquereux, Knowlton, Berry, Newberry, and Ward were from Europe or the
Arctic. In North America, the naming and sorting of these taxa took place as
these paleobotanists were caught up in the tangled web of the Laramie Problem.
The result was a snarl of poor or missing type specimens, invalid binomials,
poorly known locality data, and the assignment of incorrect botanical affinities.

As discussed in Section 4.2, stratigraphic aspects of the Laramie Problem were
largely resolved by the careful stratigraphic work of Erling Dorf (1940, 1942a,
1942b) and Roland Brown (1943, 1962) but the systematics of fossil angiosperms
remained largely a shambles. Both men used a typological approach to fossil

Figure 4.2 An example of a cleared leaf (*Tilia chingiana* Hu & Cheng). The specimen (NMNH 8269) has been bleached and stained to reveal details of venation critical for identification.

leaves and failed to address leaf architecture in a systematic manner. Brown collected extensively and came to recognize certain species in space and time, and Dorf compiled a massive compendium index of all published North America type specimens to help him unravel the contorted nomenclature of his predecessors. In the end, both men were defeated by the enormity of the problem. Brown's life work was published posthumously in 1962 through herculean efforts of Jack Wolfe, and Dorf ceased to engage in serious science after the publication of the Lance Flora in 1942.

Wolfe became Brown's successor at the US Geological Survey in 1958 and began a survey of the leaf architecture of modern angiosperms shortly thereafter. Using a technique of clearing the pigment from the leaves and staining them with a dye that concentrated in the veins, Wolfe was able to build a glass herbarium of "cleared leaves" (Figure 4.2) from vouchered herbarium sheets.

This data base allowed him to study the higher-level details of angiosperm leaf venation and search for patterns with significance for understanding the genealogical relationships of angiosperms. As a Harvard undergraduate, Wolfe was familiar with the pioneering leaf physiognomy work of I. W. Bailey and E. W. Sinnott, who had shown that angiosperm leaves contain an embedded climate signal. Specifically, they had recognized a relationship between the nature of the leaf margin (whether toothed or smooth) and mean annual temperature. Subsequent work by Ralph Chaney and his students reinforced the idea that leaf size and the presence of a drip tip was an indication of rainfall. Wolfe's research followed both of these avenues: the systematics of leaves from their vein architecture and the climate indicated by leaf physiognomy. By 1969, Wolfe's work on Alaskan fossil floras was beginning to revolutionize the study of fossil angiosperms and make legitimate connections between extant angiosperm genera and fossil ones. Dorf's only student, Leo J. Hickey, was quick to realize the potential of angiosperm leaf architecture and he began to develop a uniform nomenclature for leaf architectural features (Hickey 1973, 1979). His PhD dissertation, a study of a Paleocene–Eocene megaflora transition in North Dakota (Hickey 1977) made use of both this terminology and the leaf architectural systematics being developed by Wolfe. In 1977 Hickey joined the Smithsonian Institution and began to build a second cleared leaf collection.

At the same time, angiosperm systematics was undergoing a revolutionary synthesis. Both Arthur Cronquist at the New York Botanic Garden and Armen Tahktajan at the Komarov Botanical Institute in St. Petersburg, Russia, were working to synthesize the diversity of angiosperms. Cronquist's system (1981) and Tahktajan's (1980) shared many characteristics, and both provided a comprehensive framework for understanding the diversity of angiosperms. Hickey and Wolfe (1975) collaborated to make the first serious attempt to synthesize their knowledge of angiosperm leaf architecture with the Cronquist and Tahktajan systems. This opened the path for the rebirth of angiosperm leaf paleobotany and the potential to use the abundant fossil record to understand both the history of angiosperm evolution and the implications of that record for the interpretation of paleoclimate and extinction.

Subsequent developments in molecular phylogeny and cladistics in the 1990s have largely rewritten our understanding of angiosperm systematics, altering and improving the Cronquist/Tahktajan framework. Both the Wolfe and Hickey cleared leaf collections, now totaling more than 30 000 specimens, provide a brilliant opportunity to integrate angiosperm leaf morphology with a modern assessment of angiosperm relations.

4.5   Early responses to the Alvarez hypothesis

Within a year of publication of Alvarez et al. (1980), the K–T boundary was discovered in nonmarine strata using palynology. The boundary sections discussed by Alvarez et al. (1980) are all in marine rocks, and that circumstance presented the possibility that the anomalous concentrations of iridium (and other siderophile elements) were somehow the result of marine depositional or diagenetic processes, rather than having been derived from an extraterrestrial object striking the Earth. Charles ("Carl") Orth of the Los Alamos National Laboratory quickly assembled a team of nuclear chemists and geologists to test that possibility by seeking an iridium-bearing K–T boundary in nonmarine rocks. An important member of that team was Robert Tschudy of the US Geological Survey. Orth et al. (1981) found what they were seeking in a drill core that penetrated a sequence of coal-bearing strata at York Canyon in the Raton Basin of New Mexico (Figure 4.3). They knew from Tschudy's previous palynological analyses of coal beds in another core that the new core would encounter both uppermost Cretaceous and lowermost Paleocene rocks. In the new core, palynology was employed to bracket the K–T boundary within an interval of about one meter, and gamma-ray spectrum-analysis was utilized to pinpoint the iridium anomaly. Following that discovery, further palynological and nuclear geochemical (neutron activation) analyses of the core showed that characteristic Cretaceous pollen species present in the core abruptly disappeared precisely at the level of the peak concentration of iridium. The abrupt disappearance of certain species of pollen was easily interpreted as evidence of the extinction of the plants that had produced the pollen, because those pollen species had long been known to be present in Cretaceous rocks but not in Paleogene rocks in western North America (e.g., Tschudy 1970). Furthermore, the observation that the microstratigraphic level of the disappearance of Cretaceous pollen exactly coincided with the geochemical evidence of an extraterrestrial impact was supporting evidence for the Alvarez et al. (1980) hypothesis. It was becoming evident that an impact was the cause of K–T extinctions, not only in the marine realm but also on land.

Just one year after the Alvarez et al. paper appeared, Hickey (1981) published what purported to be a refutation of the extinction hypothesis from the perspective of paleobotany and palynology. Hickey surveyed primarily palynological data that had been published through the late 1970s and concluded that plant extinction across the K–T boundary was not catastrophic, although he did note an extinction of 75% of the flora in western North America. Hickey invoked climate change as the likely cause of plant extinctions, and also mentioned withdrawal of epeiric seas. Based on his literature survey, he asserted that plant

Figure 4.3 Map of the western United States showing sedimentary basins of Cretaceous and Paleogene age. Those containing K–T boundary sections (Raton, Denver, Powder River, and Williston) are emphasized by darker shading; see Chapters 6 and 7 for details.

extinction in western North America was not coincident with the extinction of dinosaurs in that region.

In his 1981 paper, Hickey cited as "in press" data that he set forth in a more extensive paper that was delayed in publication for three years. Hickey (1984) is the complete survey of palynological and paleobotanical literature upon which the earlier paper was based. Interestingly, the data cited indicated that the magnitude of floral change across the K–T boundary was as much as 50% in some regions of the world, but he also cited a large number of non-North American palynofloral records that appeared to indicate little or no extinction. In retrospect, the resolution of many of these records with regard to the K–T boundary is at the stage level, and most of the studies were quite preliminary as to floral composition.

In the same book *Catastrophes and Earth History – The New Uniformitarianism* that carried the paper by Hickey (1984), Tschudy (1984) published a paper specifically on the global palynological record at the K–T boundary. His paper was based on a survey of the literature published through 1977 supplemented by his own unpublished data from western North America. In commenting on the palynological record of extinction of the flora in that region also noted by Hickey (1984), Tschudy stated that the Late Cretaceous palynofloristic province had "virtually lost its identity" at the K–T boundary (Tschudy 1984, p. 333). That region aside, he acknowledged the inadequacy of the data available for addressing effects of the boundary event on plants, because no detailed evaluation of the whole flora across the K–T boundary existed. Tschudy also observed that on a global scale, palynology suffered from the absence of accurate stratigraphic control. Nonetheless, he concluded that "no prominent world-wide extinction of land plants at the end of Cretaceous time can be postulated from an examination of the pollen and spore record" (Tschudy 1984, p. 332). Clearly the paper was written years before it was published (as was Hickey's) because ironically it appeared the same year as Tschudy *et al.* (1984), the description of the fern-spore spike at the K–T boundary (see Section 5.3), which made Tschudy in effect a leading supporter of the impact theory.

Thus, early responses to the Alvarez hypothesis from the fields of paleobotany and palynology were ambiguous if not negative. However, new evidence was soon to accumulate that would change the paradigm of the fossil plant record.

4.6   New evidence accumulates

In the decade following publication of the Alvarez *et al.* (1980) impact hypothesis, the Raton Basin of Colorado and New Mexico (Figure 4.3) was a focus of studies of the palynological and paleobotanical records of the K–T extinctions. These studies were conducted in conjunction with a search for new localities at which the iridium anomaly was present in nonmarine rocks. The new evidence developed in the Raton Basin contributed supporting data for the impact hypothesis and played a role in the advancement of the hypothesis to the status of a theory. The primary publication, Orth *et al.* (1981), has already been mentioned. Confirmatory data soon followed in papers by Orth *et al.* (1982), Tschudy *et al.* (1984), and Pillmore *et al.* (1984). The same team of collaborators conducted the research and published these papers. Orth and his group at Los Alamos National Laboratory detected the iridium anomaly at new outcrop localities in the basin, Tschudy documented the palynological extinction at these localities and elaborated upon the fern-spore spike (see Section 5.3), and Pillmore led in the discovery of the new localities in the field and in their geological description. Palynology was the paleontological basis of all these investigations. By the mid 1980s, Wolfe and Upchurch (1986, 1987a) were able to conduct paleobotanical research at some of these newly discovered K–T boundary localities and outline the leaf megafossil record in the Raton Basin. The new evidence contributed by all of these studies is discussed more fully in Section 7.2.

Almost simultaneously, Jan Smit and his collaborators were making parallel discoveries at the K–T boundary in the Williston Basin (Figure 4.3), in the Hell Creek area in eastern Montana (Smit and Van der Kaars 1984, Smit *et al.* 1987), and research for a doctoral dissertation was under way, results of which were published some years later (Hotton 2002). These studies are discussed more fully in Section 6.3. As well, one of the localities at which Tschudy *et al.* (1984) found the fern-spore spike at the boundary was in the Hell Creek area, eastern Montana.

New evidence came from Canada in the mid 1980s. Lerbekmo and Coulter (1984) used the expertise of Canadian palynologist Chaitanya Singh to locate the K–T boundary in the Red Deer Valley in Alberta and the Missouri River Valley in North Dakota, correlating the stratigraphic sections in those areas, and determined that the boundary was in the upper half of magnetostratigraphic subchron C29r. Lerbekmo (1985) used palynology and magnetostratigraphy to correlate the boundary section in Alberta with a new one in the Cypress Hills area of southwestern Saskatchewan, and he reported having found the iridium anomaly in the Alberta section. Lerbekmo advocated the iridium anomaly as the most reliable physical marker of the K–T boundary. Nichols *et al.* (1986) discovered a boundary section characterized by a palynological extinction, a fern-spore spike, and an iridium anomaly in southern Saskatchewan. Jerzykiewicz and Sweet (1986) reported a palynologically defined K–T boundary in the Alberta foothills, and Lerbekmo *et al.* (1987) reported the iridium anomaly at the previously discovered K–T boundary localities in Alberta and southwestern Saskatchewan. For more detailed discussions of these Canadian localities, see Section 7.6.

Evidence continued to accumulate in ever more diverse locations in western North America in the late 1980s. Shoemaker *et al.* (1987) verified that the K–T boundary in the original York Canyon Core from the Raton Basin was within an interval of reversed polarity that, using the iridium anomaly, could be correlated with C29r. Bohor *et al.* (1987a) discovered the first K–T boundary locality in the Powder River Basin of Wyoming (Figure 4.3). In addition to a palynological extinction, fern-spore spike, and an iridium anomaly, shocked quartz was also found at this locality, as it had been in the Williston Basin in eastern Montana and in the Raton Basin (see Section 7.4). Wolfe and Upchurch (1986) compiled historical megaflora records and new data from dispersed fragments of leaf cuticle in palynological preparations from several sections containing the iridium anomaly and argued that those records supported the Alvarez hypothesis. Johnson *et al.* (1989) applied megafossil paleobotany as well as palynology to locate a new record of the boundary in the Williston Basin in North Dakota. This locality also has a small but significant iridium anomaly and shocked quartz is present (see Section 6.2). The discovery of new boundary localities continued into the 1990s and beyond, as is discussed in Chapters 6 and 7.

Thus, by the end of the 1980s, it had become overwhelmingly clear that the K–T palynological extinction was associated with an iridium anomaly on a continental scale and it was beginning to appear that an extraterrestrial impact had generated a profound and deleterious effect on terrestrial vegetation. The nature of that effect had yet to be fully explored, especially on a global scale, but more information was being derived from new K–T boundary localities. Before examining the data from nonmarine K–T boundary localities around the world, it is fitting to present a brief overview of what is currently known of the vegetation of latest Cretaceous and early Paleocene time. It is also necessary to explain in some depth that curious earliest Paleocene palynological phenomenon, the fern-spore spike.

5

Overview of latest Cretaceous and early Paleocene vegetation

5.1   Late Cretaceous vegetation

Paleobotanists divide Phanerozoic time into three eras: the Paleophytic, Mesophytic, and Cenophytic (Traverse 1988b). The Mesophytic–Cenophytic boundary lies nowhere near the Mesozoic–Cenozoic boundary. Because it is defined by the first appearance of the angiosperms, the Cenophytic begins in the earliest stages of the Cretaceous, some 60–70 million years before the K–T boundary. Angiosperm evolution was rapid in the Cretaceous and global summaries based on megafloral and palynofloral data (Crane and Lidgard 1989, Lidgard and Crane 1990, Drinnan and Crane 1990) show an equatorial origin in the Barremian followed by an explosive race to polar latitudes by the Cenomanian. Nearly all Cenomanian and post-Cenomanian megafloras are dominated by angiosperms. In terms of temperature, Wolfe and Upchurch (1987b) using leaf margin analysis documented a gentle warming from the Albian to the Santonian, followed by a gradual cooling to the early Maastrichtian, culminating with a rapid late Maastrichtian warming. Huber et al. (1995) used isotopic analysis of marine foraminifera to generate a Late Cretaceous temperature curve with similar trends but with no obvious latest Maastrichtian warming. Subsequent more detailed analysis of magnetostratigraphically calibrated marine and terrestrial records shows warming occurred within the last 500 000 years of the Cretaceous both in deep-sea cores and in terrestrial deposits (Wilf et al. 2003).

The latest pre-angiosperm Mesophytic Era (Early Cretaceous, Berriasian–Barremian) is characterized by the Wealden flora of southern England (and correlative floras in France, Belgium, Germany, Spain, and Portugal). The English portion of this flora, summarized by Watson and Alvin (1996), is composed of 4 bryophytes, 4 equisetaleans, 3 lycopods, 10 families of pterido-phytes representing about 16 species, 1 *Caytonia*, 1 pteridosperm, 12 cycads, at least 23 bennettitaleans (foliage types only), 2 ginkgophytes, 2 czekanowskialeans, and 5 families of conifers represented by about 20 species. There are two possible angiosperm species in this flora. By the Cenomanian, floras were dominated by angiosperms (e.g., the Dakota flora of Nebraska and Kansas).

The transition from Mesophytic to Cenophytic floras is thus a relatively rapid one with the classic elements of the Mesophyticum (Krassilov 1987) succumbing to the angiosperms in terms of taxonomic diversity by the beginning of the Late Cretaceous, then with respect to vegetational cover during the Late Cretaceous (Wing and Boucher 1998).

Over the course of the Late Cretaceous, many groups of non-angiosperms became relatively less important or extinct as the angiosperms became more diverse. Bennettitaleans disappeared entirely from the record by the Santonian. Other Mesophytic holdovers that survived into Campanian–Maastrichtian time include *Ginkgo*, cheirolepidiaceous conifers, and archaic cycads like *Ctenis*. Simple-leafed cycads like *Nilssonia* became increasingly less common through the Late Cretaceous, essentially disappearing in the Maastrichtian with only a few isolated occurrences in putative Paleocene and Eocene deposits of Washington State and Russia. Ferns and compound-leafed cycads continued to evolve and new forms appeared throughout the greenhouse world of the Late Cretaceous and Paleogene. Palms first appeared in the Turonian and, by the Campanian, many floras were vegetatively co-dominated by angiosperms, conifers, and ferns but taxonomically dominated by angiosperms.

Thus, just before the K–T boundary event, world floras were conspicuously Cenophytic, that is to say angiosperm-dominated, in essentially all aspects. In order to focus our survey, we will turn our attention to the flora of the Maastrichtian, the final stage of the Cretaceous. This stage represents the last 5.1 million years of the Cretaceous and is a sufficiently long span to assess floral change before and at the K–T boundary. Figures 5.1 and 5.2 are artists' recon-structions of Maastrichtian vegetation based on paleobotanical data.

Maastrichtian paleobotanical data are surprisingly sparse in a global sense. The most extensive and fossiliferous Maastrichtian terrestrial sections that contain both palynoflora and megaflora occur along the eastern flanks of the Rocky Mountain trend from the Arctic Ocean south to New Mexico; along the Amur (Heilongjiang) River of Russia and eastern China; in the Koryak Uplands and a few other areas of easternmost Russia; on the Nûgssuaq Peninsula on western Greenland; and on the South Island of New Zealand. Less extensive deposits occur on King George Island on the Antarctic Peninsula; on the Japanese island of Hokkaido; in the Pyrenees of France and Spain; in Germany

Figure 5.1 Artist's reconstruction of Maastrichtian vegetation based on paleobotanical data from the Laramie Formation at Marshall Mesa near Boulder, Colorado. Painting by Jan Vriesen. Reprinted couresy of the Denver Office of Cultural Affairs (DOCA).

Figure 5.2 Artist's reconstruction of Maastrichtian vegetation based on paleobotanical data from the Denver Formation near Strasburg, Colorado. Painting by Gary Staab. Copyright Denver Museum of Nature and Science (DMNS); reprinted by permission.

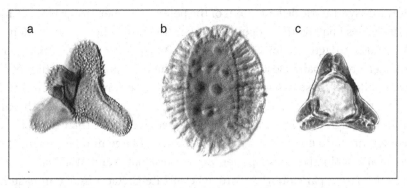

Figure 5.3 Fossil pollen representative of the *Aquilapollenites* and Normapolles palynofloristic provinces. a – *Aquilapollenites quadrilobus*, b – *Wodehouseia spinata*, c – *Extratriporopollenites* sp. (a typical species of Normapolles pollen). Specimens range from 30 to 50 micrometers in greatest dimension.

(primarily seed floras); in central India (primarily wood floras); and in the southernmost Andes of Chile and central Chubut Province in Argentina. Because of the microscopic nature of the fossils, palynological data are known from more regions, but it is notable that there is little known of either palynoflora or megaflora for the Maastrichtian of Australia, Africa, most of South America, and most of Eurasia. From this point of view, one of the most limiting aspects of the study of plants and the K–T boundary is the sparse global coverage of latest Cretaceous floras.

For this reason, global summaries have been relatively broad-brush and have drawn large regional patterns based on modest data collected at the stage level. The quality of the primary data and its stratigraphic precision are quite variable and the geographic coverage is uneven, with large gaps in the areas mentioned above outside of North America. The two primary approaches have been the delineation of paleofloristic provinces based on palynology and the reconstruction of ancient biomes based on a combination of palynology and megafloral paleobotany.

The earliest research on the definition of Late Cretaceous paleofloristic provinces based on palynology was conducted in Russia. Chlonova (1962) called attention to the existence of two groups of fossil pollen having unique morphologies that are present in northeastern Asia; they are present in western North America, as well. One group, which she designated the "unica" morphological type, includes *Aquilapollenites* (Figure 5.3a) and related genera (*Fibulapollis, Orbiculapollis*, and others) that are more commonly referred to as the Triprojectacites group or just as triprojectate pollen. Species of this group are characterized by unusual (Chlonova said "peculiar") apertures and a tendency toward asymmetric development of the poles. The other group, which Chlonova (1962) designated the "oculata" type, includes the fossil pollen genus *Wodehouseia* (Figure 5.3b). The unifying feature of this group is the occurrence of apertures in pairs close to the ends of the elongate grains. Chlonova noted that pollen of these two groups is present in Upper Cretaceous deposits of Asia from the Ural Mountains to the Far East and in western North America. Palynostratigraphic investigations conducted over more than 40 years on both continents have borne this out. The presence of large numbers of the morphologically distinctive fossil pollen of *Aquilapollenites* and *Wodehouseia* serve to characterize a palynofloristic province of Late Cretaceous age in eastern Asia and western North America commonly known as the *Aquilapollenites* Province. The *Aquilapollenites* Province disappeared as a paleofloristic entity at the end of the Maastrichtian, at least in western North America.

Coeval with the *Aquilapollenites* Province but occupying eastern North America, Europe, and western Asia was the Normapolles Province, which persisted from Late Cretaceous into Paleogene time. The characteristic angiosperm pollen species of the Normapolles Province are triporate types, most with elaborately structured apertures (Figure 5.3c). Whereas the characteristic genera of the *Aquilapollenites* Province such as *Aquilapollenites* and *Wodehouseia* represent extinct groups of plants of uncertain botanical affinity (most having became extinct at the K–T boundary), the Normapolles group of fossil pollen evidently represents plants that were ancestral to living families such as the Betulaceae and perhaps others whose pollen is characterized by three protruding pores.

Batten (1984) presented a comprehensive review of the origins of paleofloristic and palynofloristic provinces or realms through Cretaceous time that incorporates the work of Zaklinskaya (1977). Both authors discussed the *Aquilapollenites* (or Triprojectacites) Province and the Normapolles Province. Late Cretaceous marine transgressions separated these provinces in North America and western Siberia. The *Aquilapollenites* Province extended from the eastern part of the West Siberian Plain across central Siberia to the Russian Far East, northern China, and Japan to western North America. The *Aquilapollenites* Province comprises subprovinces, and districts within those, based on differences in the palynofloras such as the presence or absence of *Wodehouseia*. For example, *Wodehouseia* is not present in the Khatanga-Lena Subprovince in northern Siberia and Yakutia, but is present in the Yenisey-Amur Subprovince to the south. In western North America the palynofloras of the Canadian Arctic and Alaska resemble those of the Ust' Yenisey District of the Yenisey-Amur Subprovince, whereas those from more southerly latitudes in North America have elements in common with central Siberia and the Russian Far East (Batten 1984).

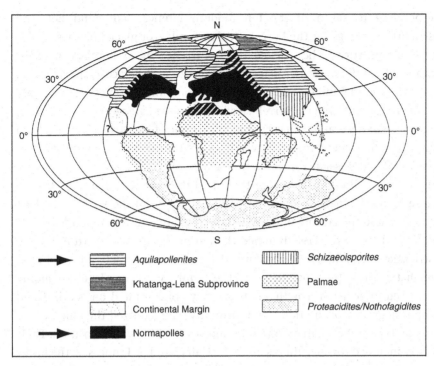

Figure 5.4 Palynofloristic provinces of the Late Cretaceous (modified from Herngreen *et al.* 1996). Reprinted by permission.

The most recent comprehensive survey of palynofloral provinces of the Cretaceous is that of Herngreen *et al.* (1996). A map presented by those authors and reproduced here in slightly modified form as Figure 5.4 diagrammatically summarizes the distribution of the *Aquilapollenites* and Normapolles provinces (see arrows) and other palynofloristic provinces of the Late Cretaceous. The Palmae Province occupied equatorial regions in Late Cretaceous time and the *Nothofagidites* Province occupied southern South America, Antarctica, New Zealand, and Australia. In northern Africa, floristic mixing of elements of the Palmae Province with those of the Normapolles Province is indicated by the diagonal black lines in Figure 5.4. Similar blurring of palynofloral province boundaries is shown for central Asia and south–central North America. In all of these areas, some palynomorph species typical of adjacent provinces may be found.

Another approach to classify Maastrichtian vegetation is the biome approach applied by Horrell (1991). He surveyed 205 megafloral and palynofloral localities worldwide and attempted to assign them to one of ten modern biome types. The modern biome types were based on precipitation, temperature, and the seasonal variability of those two factors. He recognized 12 climatic zones corresponding to seven of the ten modern biomes, but could not find Maastrichtian analogs for the boreal conifer, tundra, or glacial biomes. In the context of this survey, he tracked the occurrence of 11 families of ferns; 4 families of conifers, the Cycadaceae, Ginkgoaceae, Gnetaceae/Ephedraceae; and 22 angiosperm families, and mapped them onto his 7 biomes. At the family level, all groups surveyed by Horrell survived into the Paleocene.

The only regions that have a large enough sample of Maastrichtian megaflora and palynoflora to merit regional overviews are the Western Interior of North America and the Russian Far East. In North America, the level of terrestrial stratigraphic resolution is better for the top of the Maastrichtian than its base (because its base largely occurs in marine rocks), and in the Russian Far East both the top and the bottom of the Maastrichtian are poorly constrained.

Wolfe and Upchurch (1987b) based their summary of the North American Maastrichtian megafloras on literature dating from the 1860s to the 1940s, although they admitted that much of the family- to species-level taxonomy was in need of revision. For this reason, they concentrated on the leaf and wood physiognomy of the floras and primarily characterized the climate and vegetation type rather than defining the species-level floral composition. Their summary included the megafloras of the McNairy Sand of Tennessee; the lower Atanekerdluk Formation of west Greenland; the Raton and Vermejo formations of the Raton Basin of Colorado and New Mexico; the Laramie and Denver formations of Colorado; the Hell Creek, Lance, and Fox Hills formations of the Powder River and Williston basins of Wyoming, Montana, and the Dakotas; and the Brazeau Formation of Alberta. They concluded that the Maastrichtian floras grew in a subhumid climate that at the end of the Maastrichtian was as warm as the Late Cretaceous thermal maximum of the Santonian. In a subsequent paper, Upchurch and Wolfe (1993) used the historical data to compare the floristic composition of Maastrichtian megafloras from Saskatchewan, Montana, Wyoming, Colorado, New Mexico, and Tennessee. Angiosperms account for 79–95% of the species diversity. Conifers account for 0–10%, and pteridophytes are similarly low at 0–14%. The only cycad recorded in these floras is *Nilssonia*, and *Ginkgo* is present in some floras. On a regional to continental scale, the Maastrichtian flora of central North America is clearly dominated by angiosperms.

Johnson (2002) approached the issue of Maastrichtian megaflora diversity by making large collections from the uppermost Maastrichtian Hell Creek Formation in North Dakota. He collected 9627 specimens from 106 localities and assigned them to one of 328 megafloral morphotypes. Based on leaf taxa from this data set, he characterized the late Maastrichtian megaflora as 92.1% angiosperm, 3.6% pteridophyte, 3.3% conifer, 0.3% cycad (only *Nilssonia*), 0.3%

Ginkgo, and 0.3% bryophyte. These data supported the continent-wide pattern observed in the historical data by Upchurch and Wolfe (1993). In terms of taxonomy, the Hell Creek flora has a few rare ferns, one cycad, one ginkgo, and conifers representing the Taxodiaceae, Cupressaceae, and Araucariaceae. The identified angiosperms are dominated by the Lauraceae, Platanaceae, Magnoliaceae, Cercidiphyllaceae, Araceae, and Arecaceae (Palmae). A number of other extant families that are present in the flora include the Berberidaceae, Ranunculaceae, Cannabaceae, Urticaceae, Ulmaceae, Malvaceae, Rosaceae, Rhamnaceae, and Cornaceae. A large proportion of the flora is composed of dicotyledonous angiosperms that have not been identified to subclass. Thus, the overall affinities of the flora still remain in need of systematic study. One interesting characteristic of the late Maastrichtian flora of North Dakota is the high percentage of angiosperms with pinnately or palmately lobed leaves. At some localities, up to 30% of the species have lobed leaves (see Section 6.2).

The Maastrichtian megaflora of the Russian Far East has been summarized by Krassilov (2003), who focused on megafloras from the area along the Amur River, the Koryak Uplands, and the west coast of Sakhalin Island (see Section 8.5). The correlation of these Russian sections to the geological time scale in general, and to the Maastrichtian stage specifically, is not yet well constrained by geochronology, magnetostratigraphy, or globally accepted biostratigraphy. Nonetheless, these megafloras are associated with palynofloras of the *Aquilapollenites* Province and they bear a striking resemblance to North American Maastrichtian megafloras.

Based on the sample of Late Cretaceous megaflora from North America and from the Russian Far East, two patterns become apparent in the Northern Hemisphere: angiosperms are the overwhelming dominant and are extremely diverse in the Maastrichtian, and conifers are vegetatively more common at higher latitudes. Essentially no Maastrichtian megafloras are known from equatorial regions, but this pattern is beginning to change with the work of Carlos Jaramillo and students in Colombia (e.g., Jaramillo and De la Parra 2006).

Southern Hemisphere Maastrichtian megafloras are known primarily from the South Island of New Zealand (Pole 1992, Kennedy *et al.* 2002), the Antarctic Peninsula (Dutra and Batten 2000), and Patagonia (Yabe *et al.* 2006). All of these areas were at high paleolatitudes in Maastrichtian time. These floras are modestly well known and are composed of podocarpaceous and araucarian conifers, taeniopterid cycadophytes, pteridophytes, and a variety of angiosperms including Betulaceae, Lauraceae, Sterculiaceae(?), Proteaceae, and large-leaved forms of *Nothofagus*. An interesting pattern evident in these floras is the appearance of lobed angiosperms. Lobed angiosperms, associated with the deciduous habit in temperate mesic climates, are effectively absent from the extant, mesic, mixed conifer and broad-leafed evergreen forests of New Zealand and Chile. The presence of lobed leaves in the Maastrichtian floras of New Zealand, Antarctica, and Patagonia suggests that these floras may have been deciduous.

In summary, Maastrichtian megafloras are known from mid to high latitudes in both hemispheres, but are essentially unknown at equatorial latitudes. Broad floristic and biomal patterns can be drawn by palynology, but these are limited in their botanical resolution. At a global scale, the transition from the Mesophytic to the Cenophytic was complete by the Maastrichtian and floral change at the K–T boundary occurred in a world where angiosperms represented both floristic and vegetational dominance.

5.2   Early Paleocene vegetation

Paleocene vegetation is best known from the Rocky Mountains and Great Plains of North America, northeastern Asia, and Greenland. Isolated Paleocene sites have been discovered or described in Europe, the Canadian Arctic, Antarctica, New Zealand, India, Colombia, Chile, and Argentina. Africa and Australia are poorly known, with essentially no Paleocene megafloras. The North American Paleocene is so well sampled from nearly a thousand fossil localities that knowledge of Paleocene megafloras is probably an order of magnitude greater than that of Maastrichtian megafloras from the same regions. Probably the only benefit of this great data disparity is that the dense sampling of the Paleocene relative to the Cretaceous provides a situation where survivors of the K–T boundary event are likely to be well sampled and K–T boundary last-appearance data are likely to be reliable.

The most extensive exposures of Paleocene terrestrial strata in the world are present in the Laramide basins of New Mexico, Colorado, Utah, Wyoming, Montana, North Dakota, and South Dakota (Figure 4.3), and in Saskatchewan and Alberta. Aggressive fieldwork and collection of fossil plants in these basins began with the US Geological and Geographic Surveys of the Territories (later by the US Geological Survey) and the Geological Survey of Canada and paleobotanists working for these agencies produced a series of monographs. Authors included John Strong Newberry, Leo Lesquereux, Lester Ward, Frank Knowlton (1922, 1930), Edward Berry, Walter Bell, and Roland Brown (1943, 1962). Brown's (1962) posthumous summary of the United States Paleocene, the "Paleocene flora of the Rocky Mountains and Great Plains," was a comprehensive survey of 470 fossil plant localities from 25 formations in seven states based on nearly 30 years of field and museum research. Brown, by his own admission, was a taxonomic lumper who treated fossil entities much more conservatively than did his predecessors. His composite flora included 169 species including

126 dicotyledonous angiosperms, 13 monocotyledonous angiosperms, 13 ferns, 7 conifers, 6 bryophytes, 1 *Gingko*, 1 cycad, 1 *Equisetum*, and 1 lycopod. He established that the Paleocene flora included the first occurrence of many extant angiosperm genera in the Salicaceae (*Salix*), Juglandaceae (*Carya, Juglans, Pterocarya*), Betulaceae (*Betula, Corylus*), Fagaceae (*Castanea, Quercus*), Ulmaceae (*Celtis, Ulmus*), Lauraceae (*Sassafras*), Sapindaceae (*Acer*), and Cornaceae (*Cornus, Nyssa*). Other genera that he identified to extant genera are almost certainly different elements that bear a superficial resemblance to extant taxa. These misattributions include Moraceae (*Artocarpus, Ficus, Morus*), Rosaceae (*Prunus*), Aquifoliaceae (*Ilex*), and Vitaceae (*Ampelopsis, Cissus, Parthenocissus, Vitis*). Brown focused much of his energy on the K–T boundary and as a result he was very interested in the stratigraphic distribution of plant species. Nonetheless, Brown's summary did not treat this issue as anything more than a general matter, and the precise stratigraphic range of Paleocene species had to wait for subsequent researchers including Peter Crane, Leo Hickey, Kirk Johnson, Steven Manchester, Elisabeth McIver, Peter Wilf, Scott Wing, and Jack Wolfe, who greatly increased the number of known Paleocene megafloral localities.

The geographic and temporal density of Paleocene megafloral sites in these rocks provide the opportunity to assess the regional vegetation over a ten-million-year period both in terms of latitude and with respect to local sedimentological facies and position within depositional basins. Collectively, the Paleocene megaflora of the Rocky Mountains is an impressive record of nearly a thousand localities spanning a contiguous region of more than a million square kilometers. It is arguably the largest and most continuous fossil floral sequence of any age. The portion of this sequence exposed in the Powder River Basin of Wyoming contains some of the world's thickest coal seams and the largest coal reserves in North America.

The abundance of Paleocene plant sites presented an opportunity for isolated organs from the same plant species to be associated, based on multiple co-occurrences. Steve Manchester from the University of Florida has aggressively pursued this approach and he and his colleagues have successfully reconstructed several plants based on this methodology (e.g., Manchester *et al.* 1999). Based on this work, it is clear that the Paleocene contained an abundance of Cornaceae, Juglandaceae, Cercidiphyllaceae, Betulaceae, Ulmaceae, and Hippocastanaceae. These families are represented by both extant and extinct genera.

Hickey (1977) published the flora of uppermost Paleocene Golden Valley Formation of North Dakota and brought an understanding of leaf architecture, the importance of sedimentary facies, and temporal resolution to the study of Paleocene floras. In 1980, he published a summary of the Paleocene flora of the

Figure 5.5 Artist's reconstruction of early Paleocene vegetation based on paleo-botanical data from the excavation for the Denver International Airport. Painting by Donna Braginetz, copyright DMNS; reprinted by permission.

Clarks Fork Basin, a northern extension of the Bighorn Basin in Wyoming, based on more than 10 000 specimens from 66 localities distributed through the Fort Union Formation, which is up to 3000 m thick. The Clarks Fork Basin and the adjacent Bighorn Basin is the only region in the Western Interior where all four of the Paleocene land mammal stages (Puercan, Torrejonian, Tiffanian, Clarkforkian) are known in superposition. Hickey's localities were tied to mammal localities and thus represented the first complete sequence of Paleocene floras. He recognized 74 species that he used to define four mega-floral zones. The lowest zone roughly corresponds to the Puercan land mammal stage, radiometrically dated as the first million years of the Paleocene (Swisher *et al.* 1993).

Johnson and Hickey (1990) subsequently recognized that this basal Paleocene flora was also present in North Dakota, Wyoming, and Montana and recorded it at 34 additional localities. They gave this megafloral zone the informal name Fort Union I or FUI. McIver and Basinger (1993) monographed an FUI assemblage from Saskatchewan. Additional localities in North Dakota discussed in Johnson (2002) refined the content of FUI, and Barclay *et al.* (2003) documented the FUI flora in the Denver Basin (Figure 5.5). The composition of the basal Paleocene

Table 5.1 *Megafloral species common in the basal Paleocene FUI assemblage*

Allantodiopsis erosa (Lesquereux) Knowlton and Maxon 1919
Woodwardia gravida Hickey 1977
Onoclea hesperia Brown 1962
Ginkgo adiantoides (Unger) Heer 1870
Metasequoia occidentalis (Newberry) Chaney 1951
Glyptostrobus europaeus (Brongniart) Heer 1855
Mesocyparis borealis McIver and Basinger 1987
Fokieniopsis catenulata (Bell) McIver and Basinger 1993
Porosia verrucosa (Lesquereux) Hickey 1977
Limnobiophyllum scutatum (Dawson) Krassilov 1976
Nelumbium montanum Brown 1962
Paleonelumbo macroloba Knowlton 1930
Harmsia hydrocotyloidea McIver and Basinger 1993
Quereuxia angulata (Newberry) Khrystofovich 1953
Paranymphaea crassifolia (Newberry) Berry 1935
Cornophyllum newberryi (Hollick) McIver and Basinger 1993
"Populus" nebrascensis (Newberry) Lesquereux 1888
Ziziphoides flabella (Newberry) Crane *et al.* 1991
Cercidiphyllum genetrix (Newberry) Hickey 1977
Browneia serrata (Newberry) Manchester and Hickey 2007
Platanites marginata (Lesquereux) Johnson 1996
Platanus raynoldsii Newberry 1868
Cissites panduratus Knowlton 1917
Penosphyllum cordatum (Ward) Hickey 1977
"Zizyphus" fibrillosus (Lesquereux) Lesquereux 1878
Beringiaphyllum cupanioides (Newberry) Manchester, Crane, and Golovneva 1999
Celtis aspera (Newberry) Manchester, Akhmetiev, and Kodrul 2002
Averrhoites affinis (Newberry) Hickey 1977

FUI megaflora is presented in Table 5.1. It is important to note that many FUI megafloral taxa occur over a paleogeographic range exceeding 10 degrees of latitude.

Despite more than a century of intensive sampling of the Paleocene Rocky Mountain floras and a broad understanding of their composition, the 1994 discovery of a hyperdiverse, 64-million-year-old tropical rainforest megaflora from the western margin of the Denver Basin in Colorado (Johnson and Ellis 2002) suggested that the Paleocene megafloral record, however extensive, was still incomplete (Figure 5.6). Compositionally similar but an order of magnitude more diverse than other Paleocene floras, the Castle Rock site presented a conundrum that remains unresolved today. This is discussed in Section 7.3.

Figure 5.6 Artist's reconstruction of the 63.8 Ma tropical rainforest vegetation based on paleobotanical data from the upper part of the Denver Formation in Castle Rock, Colorado. Painting by Jan Vriesen. Copyright DMNS; reprinted by permission.

In the Paleocene, the palynofloristic provinces that had existed in the latest Cretaceous were significantly altered. Most dramatically, the *Aquilapollenites* Province in western North America ceased to exist (Tschudy 1984). In western North America, up to 45% of the Maastrichtian palynoflora – primarily angiosperm pollen – disappeared from the region (Nichols 1990, Nichols and Fleming 1990, Nichols *et al.* 1990). All but one of the uppermost Maastrichtian species of the genus *Aquilapollenites* vanished abruptly at the K–T boundary. The lone surviving species, *Aquilapollenites reticulatus* (also known as *A. mtchedlishvilii*), is present in the lowermost Paleocene as well as in the upper Maastrichtian, but it disappeared in the middle Paleocene. In well-studied sections in the Williston Basin of North Dakota and Montana, about 30% of the palynoflora including species of *Aquilapollenites* and many others became extinct within a few centimeters of the K–T boundary (Hotton 2002, Nichols and Johnson 2002), and an additional 20–30% underwent a statistically significant decline in abundance (Hotton 2002).

Early Paleocene megafloras from northeastern Asia are represented by significant localities on both the Russian and Chinese sides of the Amur River (Krassilov 1976, Kodrul 2004) and the Koryak Uplands (Golovneva 1994a, b).

These floras contain a great number of genera in common with Paleocene localities in the Western Interior of North America. Some of these taxa (e.g., *Celtis aspera*, *"Ampelopsis" acerifolia*, *Beringiaphyllum cupanioides*) are not known from the Cretaceous on either continent, suggesting a continuity of connection across the Bering Land Bridge into the early Paleocene. Unlike the Paleocene megafloras from the central part of North America, the Russian and Chinese Paleocene megafloras have not been independently dated and the temporal ranges of specific taxa are poorly known.

In the Siberian region, palynofloristic changes from the Maastrichtian to the Paleocene were profound. Bratseva (1969) summarized data that indicate that in Danian (Paleocene) time, taxa that had been subordinate in the Maastrichtian became widespread, and ancestral members of the modern families Myricaceae, Betulaceae, and Juglandaceae became dominant.

In western North America, lower Paleocene palynofloras are severely reduced in diversity in comparison with those of the Maastrichtian (Hotton 2002, Nichols and Johnson 2002). In these assemblages, a few species tend to dominate numerically. Low diversity in lower Paleocene palynomorph assemblages is also reported in the Neotropics of the Western Hemisphere (Jaramillo *et al.* 2006). The most extreme examples of low-diversity palynologic assemblages in the lowermost Paleocene constitute the fern-spore spike at the impactite-level of stratigraphic resolution (see Section 2.3); these assemblages, which are discussed in the next section, may consist almost entirely of a single species of fern spore. At the stage level of resolution, palynofloras from the lower Paleocene are characterized more by the absence of Maastrichtian species than by the presence of new Paleocene taxa, which appeared gradually later (Nichols and Johnson 2002, Nichols 2003). Based on palynologic evidence, the family Juglandaceae, which has a negligible record in the Maastrichtian, radiated through Paleocene time, and its many species now serve as the basis of palynostratigraphic subdivision of the stage (Nichols 2003). These and many other new species that characterize Paleocene assemblages tend to have close affinity with living taxa.

In contrast to the *Aquilapollenites* Province in western North America, the Normapolles Province of the Late Cretaceous persisted through the Paleogene in eastern North America, Europe, and western Asia, although waning in numbers of taxa (Batten 1981). The late Paleogene decline and eventual disappearance of the Normapolles group of pollen that defines the palynofloristic province seems to be a result of the gradual evolution of the pollen-producing plants into modern taxa of the order Juglandales (Traverse 1988b). A contributing factor to the dissolution of the province may have been the Paleocene withdrawal of the epicontinental seaways from North America and Asia, seaways that had effectively separated the Normapolles and *Aquilapollenites* provinces in Cretaceous time (Batten 1981). This resulted in geographic mixing of evolving taxa in plant communities and biomes across the Northern Hemisphere.

Southern Hemisphere Paleocene megafloras are known primarily from the South Island of New Zealand, the Antarctic Peninsula, and southern Patagonia. Sampling is meager and stratigraphic relations are still too poorly known for us to understand how these floras differ from Upper Cretaceous floras from the same regions.

In general, early Paleocene floras grew in a warm greenhouse world that was forested from pole to pole. Paleocene forests contained the earliest examples of many extant genera. As with the Maastrichtian, very little is known of Paleocene floras from the equatorial regions, and despite the presence of early Paleocene tropical rainforests at mid latitudes, there is little to no evidence for them at low latitudes.

5.3   The fern-spore spike

A unique feature of earliest Paleocene vegetation is an anomalous concentration of fern spores just above the level of palynological extinction. This anomalous palynological assemblage is known as the fern-spore spike. It is evidence that plant communities of the earliest Paleocene were pioneer communities numerically dominated by ferns, in most places of a single species. An artist's conception is on the cover of this book.

The first report of this exceptional palynological assemblage was that of Orth *et al.* (1981). In addition to determining that a palynological extinction horizon and the peak of an iridium anomaly were at the same stratigraphic level, palynological analysis of a drill core that penetrated the K–T boundary revealed another startling phenomenon: just above the extinction level, the ratio of angiosperm pollen to fern spores dropped abruptly and radically (Figure 5.7). A normal ratio of pollen to spores was reestablished several centimeters above the K–T boundary. In its original form, it was described as a change in the ratio of angiosperm pollen to fern spores, but since that time it has been depicted as an increase in the relative abundance of fern spores – a spike in abundance of fern spores.

The implications of the fern-spore spike were not understood at the time it was first observed by R.H. Tschudy (in Orth *et al.* 1981), but following his discovery in the drill core from New Mexico, Tschudy *et al.* (1984) analyzed samples from K–T boundary outcrop localities in Colorado and Montana. He too observed the same phenomenon just above the palynological extinction

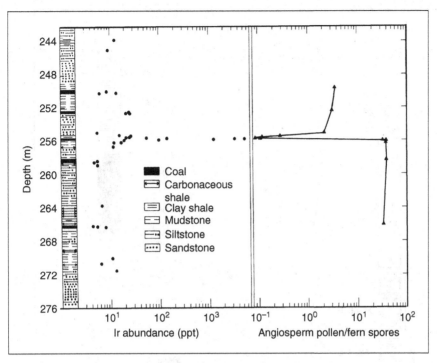

Figure 5.7 Iridium concentrations (black dots) and ratios of angiosperm pollen to fern spores (triangles) in the York Canyon Core; major deflections mark the K–T boundary (from Orth *et al.* 1981). Reprinted by permission.

horizon at the outcrop localities (Figure 5.8). Realizing that data points as far apart as New Mexico and Montana were evidence of a continent-wide event, Tschudy made an insightful interpretation about the plant communities that flourished just after the K–T extinction event. He had observed that not only were palynological assemblages of the fern-spore spike overwhelmingly dominated by spores of ferns, but that almost all of the spores were of a single species. This observation conjured up an image of a landscape almost completely covered by a single species of fern, a pioneer species on a terrain laid waste by the extinction event. The first plants to reinhabit the devastated, post-extinction landscape were ferns. Ferns have the ability to regenerate quickly from rhizomes and quickly produce vast numbers of reproductive bodies (spores); and perhaps more importantly, as non-seed plants, they easily reproduce themselves from spores alone.

Tschudy found a modern analogue for this pioneer plant community in descriptions of the volcanic island of Krakatau (Krakatoa), Indonesia, which had been essentially wiped clean of vegetation by a cataclysmic explosion in 1883. Richards (1952) published an account of the recolonization of Krakatau by

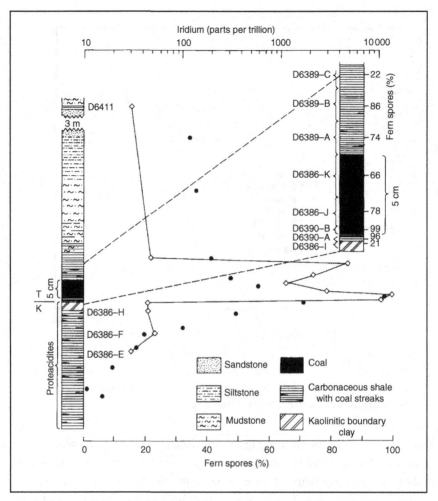

Figure 5.8 Diagram showing variations in iridium concentration (black dots) and fern-spore percentages (open diamonds) at the Starkville North K–T boundary locality in the Raton Basin, Colorado (from Tschudy *et al.* 1984). Reprinted by permission.

plants. The earliest visitors to the caldera found no plants living on it. A botanist arriving in 1886 found that some plants had returned and was impressed that most of them were ferns. After a few years, immigrant species from nearby islands evidently reestablished the former, diverse communities, although Richards noted that seeds, spores, and underground parts of plants could also have survived in ravines on the devastated island. Such areas could have served as refugia from which the communities were reestablished. Thus, the scenario suggested by the presence of fern-spore spikes just above the K–T boundary at widely separated locations contributed much to an understanding of the effects of the terminal Cretaceous event on plant communities.

Fleming and Nichols (1990) codified the nature of fern-spore spike assemblages at the K–T boundary and their differences from other palynological assemblages rich in fern spores found elsewhere in the stratigraphic record. They proposed that for an assemblage to qualify as a fern-spore abundance anomaly or spike, its composition must be 70–100% fern spores of a single species occurring within an interval 0–15 cm above the K–T boundary (Figure 5.9). Fleming and Nichols accepted Tschudy's interpretation of the significance of such assemblages – that they represent surviving species that ubiquitously colonized earliest Paleocene landscapes.

Robert Tschudy's original interpretation of the nature and significance of fern-spore spikes did not go unchallenged. Arthur Sweet of the Geological Survey of Canada and his colleagues at first questioned the existence of fern-spore spikes because they observed none at K–T boundary localities in Canada. Eventually they detected fern-spore spikes at Canadian localities using micro-stratigraphic sampling and they acknowledged the occurrence of anomalous spore percentages, but disputed the interpretation of the fern spores as representing colonizing vegetation on a devastated landscape.

Sweet and Braman (1992) described occurrences and relative abundances of groups of various palynomorphs (pollen, spores, and algal cysts) through stratigraphic intervals at twelve Canadian K–T boundary localities. They reported geographic differences in which fern taxa established numerical dominance shortly after the extinction event, and they expanded the concept by identifying three species of angiosperms that substituted as pioneer species (they preferred to call them "opportunistic species") in place of ferns at some localities (Figure 5.10). They asserted that the particular species (fern or angiosperm) that briefly assumed dominance in the local flora was determined by which species were prevalent in the local area just prior to the extinction event, which appeared to be related to paleolatitude. This is an interesting expansion of the concept of pioneer species flourishing after the extinction, but with regard to angiosperms – which reproduce via seeds instead of spores – it has less theoretical support from botany than does the original explanation of the phenomenon published by Tschudy et al. (1984). Nonetheless, the work of Sweet and Braman (1992) shows a pattern of paleolatitudinal variation in which species are the first to colonize the local landscape following the extinction event. Sweet and Braman noted that ferns tended to be the pioneer or "opportunistic" species in the southern part of the Western Interior region and a few species of angiosperms assumed that role in the northern part.

A final viewpoint offered on post-extinction, fern-dominated plant communities requires comment. In their paper on the megaflora of the Raton Basin discussed in Section 7.2, Wolfe and Upchurch (1987a) tended to shift emphasis

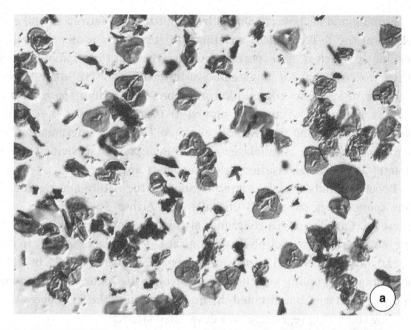

Figure 5.9 Views of the fern-spore spike. a – low-power magnification, b – high-power magnification. All specimens are *Cyathidites diaphana* except the bean-shaped spore (*Laevigatosporites* sp.) at center right in the low-power view. Both the high abundance and low diversity of spores characteristic of the fern-spore spike are illustrated. Specimens of *C. diaphana* are 25–30 micrometers in diameter.

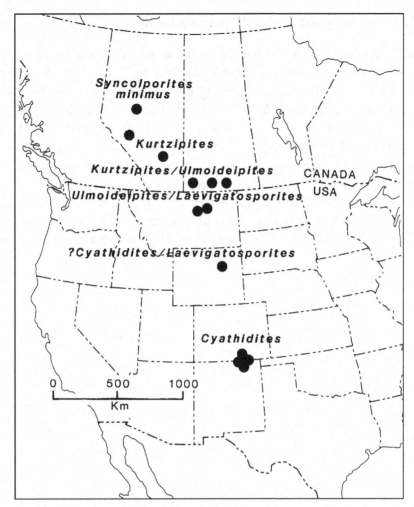

Figure 5.10 Map showing latitudinal variation in dominant taxa of the fern-spore spikes and angiosperm-pollen-dominated assemblages just above the K–T boundary (from Sweet and Braman 1992). Reprinted by permission of Elsevier.

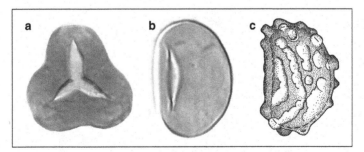

Figure 5.11 Fern spores. a – *Cyathidites diaphana*, b – *Laevigatosporites* sp., c – *Stenochlaena tenuifolia*. *Cyathidites* and *Laevigatosporites* characterize fern-spore spikes from New Mexico to Saskatchewan; spores of *Stenochlaena* bear no resemblance to either, despite its having been identified as the fern spike genus on the basis of megafossils by Wolfe and Upchurch (1987a). Illustration of *Stenochlaena* from Erdtman (1957). Specimens "a" and "b" 30 micrometers, "c" 45 micrometers.

of the nature of the fern-spore spike interval toward a megafloral perspective. They reported a dominance of fern fronds at three Paleocene localities and referred to it as a "fern spike" interval. They stated that the fronds they collected represent an extinct genus related to the extant fern genus *Stenochlaena*, a primary colonizer in the modern vegetation of Indomalaya and Africa. Palynological records from the fern-spore-spike interval do not support this interpretation, however appealing it might be. Two species of fern spores representing two different fern genera, *Cyathidites* and *Laevigatosporites* (Figure 5.11a, b), characterize the fern-spore spikes in the Raton Basin and elsewhere in North America, and neither of these genera includes species that resemble the distinctive morphology of spores of *Stenochlaena* (Figure 5.11c).

PART II REGIONAL CASE STUDIES

6

Williston Basin – the most complete K–T sections known

6.1 Overview

The uppermost Maastrichtian and lowermost Paleocene rocks in the Williston Basin of western North Dakota, northwestern South Dakota, and eastern Montana contain the best exposed and most studied nonmarine record of the terminal Cretaceous event in the world. This region contains 41 (39%) of the known terrestrial K–T boundary sections (see Table 2.1 and Appendix). The Williston Basin is a large structural depression occupying much of North Dakota and parts of South Dakota, Montana (Figure 6.1), and it extends into southern Saskatchewan. The stratigraphic units in which the K–T boundary is preserved are the Hell Creek Formation, which for the most part is Maastrichtian in age, and the Fort Union Formation, which for the most part is Paleocene in age (Figure 6.2). The Hell Creek Formation is composed of fluvial sandstone, mudstone, and claystone. It includes rare, thin lignite beds and minor marine units. It was deposited at the edge of the Western Interior seaway (Roberts and Kirschbaum 1995, Johnson *et al.* 2002). The Fort Union Formation is composed of extensive lignite and carbonaceous shale beds, variegated mudstone of lacustrine origin, and sandstone largely of crevasse-splay origin. The facies change that marks the formational contact was caused by rising sea level during a temporary re-advance of the Western Interior seaway in early Paleocene time known as the Cannonball Sea, a Paleocene remnant of the Cretaceous seaway represented by the Cannonball Member of the Fort Union Formation (Figure 6.2). The K–T boundary is stratigraphically near (0–3 m), and in some places precisely coincident with, the Hell Creek–Fort Union formational contact.

These formations preserve a wealth of plant megafossils and palynomorphs as well as a rich variety of invertebrate and vertebrate fossils. Outcrops are good

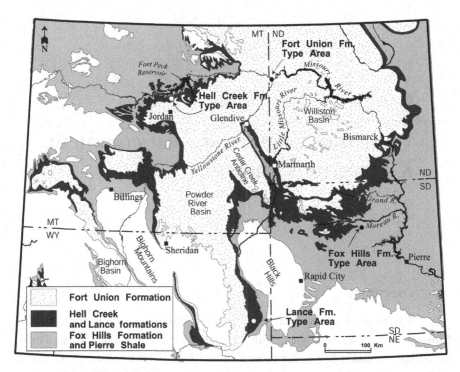

Figure 6.1 Outline map of the Williston Basin in western North Dakota, eastern Montana, and northwestern South Dakota and the Powder River Basin in southeastern Montana and northeastern Wyoming. The northernmost part of the Williston Basin extends into adjacent southern Canada. Williston Basin margin is defined by the Hell Creek–Fort Union Formation contact and Powder River Basin margin is defined by the Lance–Fort Union Formation contact. These formation contacts approximate the K–T boundary (from Johnson *et al.* 2002). Reprinted by permission.

to excellent as the climate of the central Great Plains is arid and the soft bedrock weathers to form extensive but discontinuous areas of badlands dissected by tributaries of the Missouri River (Figure 6.3). The Williston Basin is dinosaur country, from which the first known skeleton of *Tyrannosaurus rex* came in 1905 and where, 85 years later, perhaps the most famous *T. rex*, "Sue," was discovered. The region has yielded numerous skeletons of *Triceratops horridus* now on display in natural history museums around the world, and it is from this region that most of the iconography of vertebrate extinction at the K–T boundary has emerged. See Russell and Manabe (2002) and references therein for more about the dinosaur fauna of the Hell Creek Formation.

In Maastrichtian time, the area of the Williston Basin was a broad alluvial coastal plain sloping gently eastward toward the Western Interior seaway. Medium-sized rivers meandered across the coastal plain. Mires and swamps

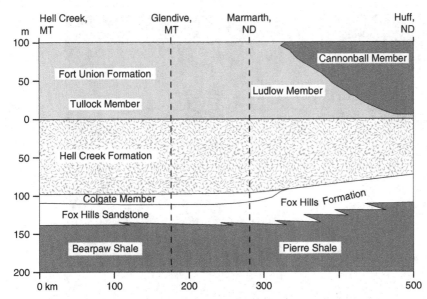

Figure 6.2 Stratigraphic nomenclature for the uppermost Cretaceous and lower-most Paleogene of the Williston Basin in eastern Montana and western and south-central North Dakota. The Hell Creek Formation and Tullock and Ludlow Members of the Fort Union Formation are nonmarine; the Bearpaw Shale and Pierre Shale and the Cannonball Member of the Fort Union Formation are marine (from Johnson *et al.* 2002). Reprinted by permission.

were rare, but tended to develop as the water table rose during minor incursions of the sea. A diverse forest dominated by angiosperms covered the area. The flora comprised hundreds of species and included both familiar groups such as palms and a variety of lobe-leafed species, many of which are unknown today in forests anywhere in the world. The Maastrichtian climate was warm, especially during the last million years. Toward the end of Maastrichtian time, the warmest temperatures promoted an influx of thermophyllic (warmth-loving) plant species into the basin; these additions to the flora served to bring species diversity to a peak. At 65.5 Ma, at the K–T boundary, the vegetation of the Williston Basin was abruptly, profoundly, and permanently changed. The destruction of the flora involved extinction of all of the dominant plants of the Maastrichtian.

The Paleocene climate did not change radically from what had prevailed during Maastrichtian time. Mean annual temperatures remained about the same as they had been, as indicated by the persistence of thermophyllic species such as palms. The terrain changed somewhat in early Paleocene time, largely in response to a major re-advance of the Western Interior seaway in the form of the Cannonball Sea. Rivers were effectively dammed, the water table rose across the region developing extensive mires in which peat accumulated. A different flora

Figure 6.3 Outcrops of the Hell Creek and Fort Union formations near Marmarth in the Williston Basin of southwestern North Dakota. a – view of Bobcat Butte showing badlands of the Hell Creek Formation in the foreground and the lowermost part of the Fort Union Formation at the top of the butte in the distance. b – view of Pretty Butte, which is composed of both Hell Creek and Fort Union strata; the black line approximates the position of the formation contact and the K–T boundary.

replaced the dry-land forests of the Maastrichtian, which largely had been destroyed by the terminal Cretaceous event. There was no recovery of the former species richness in the Williston Basin. The flora remained relatively static and depauperate for at least the first million years or so of the Paleocene. At first, ferns and peat moss were the most common plants. Later, angiosperms reclaimed dominance, but although palms persisted, most of the species of the low-diversity Paleocene forests were those that had been extremely rare in the region during the Maastrichtian. Toward the end of the Paleocene, forests were composed largely of cupressaceous-taxodiaceous conifers. These swamp-loving trees contributed much to the accumulation of the peat deposits that would eventually form thick coal beds.

6.2   North Dakota

Most of what is known about plants and the K–T boundary in the Williston Basin region, as summarized in this section, comes specifically from the Marmarth area in the southwestern corner of North Dakota (Figure 6.4). It derives from our own research, Johnson on paleobotany and Nichols on palynology, mostly in collaborative studies on the Maastrichtian Hell Creek Formation and Paleocene lowermost part of the Fort Union Formation. The flora – both megaflora and palynoflora – of the Maastrichtian and Paleocene in this area has been thoroughly sampled and extensively studied, although much systematic work remains to be done on the megaflora. Paleobotanical and palynological results from this area incorporate physical evidence of the K–T boundary (the iridium abundance anomaly, shock-metamorphosed minerals, and spherules). Many of the paleobotanical and palynological collections are associated with intensively studied vertebrate fossils, and many of the measured sections from which the fossils came have also been analyzed paleomagnetically and geochemically. A comprehensive summary of all recent studies in the area is the *Geological Society of America Special Paper* 361 (Hartman *et al.* 2002).

The plant fossil record is based on more than 160 leaf quarries and more than 350 individual microfossil samples (see Figure 6.5). Unusual for a paleobotanical study is the large number of megafossil specimens collected, more than 22 000. Plant microfossils, being so much smaller, are generally easier to collect in large numbers, yet few other studies in the field of palynology have involved an actual survey of some 700 000 specimens, as is true of the North Dakota K–T boundary study. The megafloral analysis is based on 380 morphotypes (species equivalents), and the palynofloral analysis involves about 110 systematically investigated taxa. Southwestern North Dakota is not where the first investigations of floral changes across the K–T boundary were conducted (see Section 7.2), but it is the area that has seen the most intensive study, and from which the most complete record of these changes comes. It is instructive to review the historical highlights of the development of the data and interpretations, focusing on work that was done in the 1980s in response to the Alvarez challenge.

The first published report on the K–T boundary from the area is that of Johnson *et al.* (1989), which summarized a collaborative study by paleontologists and nuclear chemists at the site named Pyramid Butte. That locality yielded all the essential elements of a well-documented K–T boundary locality in

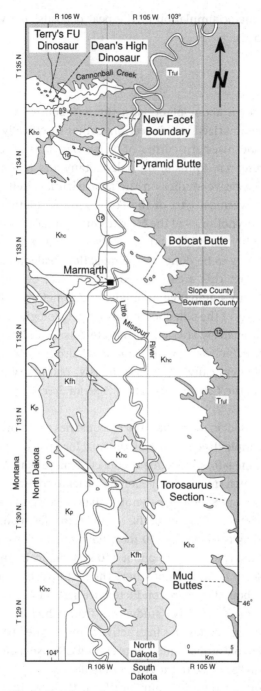

Figure 6.4 Geological map of the area in the vicinity of the town of Marmarth, southwestern North Dakota, showing locations of K–T boundary sites discussed in the text (modified from Nichols and Johnson 2002). T – township, R – range, Kp – Pierre Shale, Kfh – Fox Hills Formation, Khc – Hell Creek Formation, Tful – Ludlow Member of the Fort Union Formation. Reprinted by permission.

Figure 6.5 A high-resolution sampling of the Hell Creek–Fort Union formation contact in North Dakota. In this case, the K–T boundary is located nearly 2 m above the base of the lignite bed that marks the formation contact, which is where the people are seated at the trench.

continental rocks: megaflora, palynoflora, iridium, and shocked quartz (for a comparison of K–T boundary localities based on definitive criteria, see Table 2.1). Leaf quarries that yielded Maastrichtian species below and Paleocene species above bracket the boundary level. Palynological collections made at a

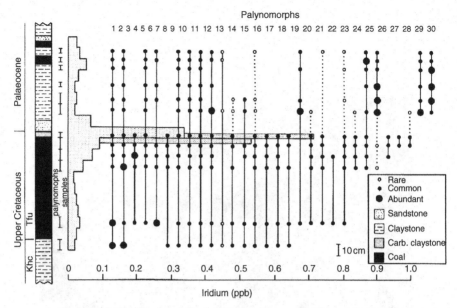

Figure 6.6 Distribution and relative abundances of iridium and the most common palynomorphs in the K–T boundary section at Pyramid Butte (from Johnson *et al.* 1989). Khc, Hell Creek Formation; Tfu, Fort Union Formation. Numbers across top designate species of spores and pollen not individually named here: 1–3, bryophytes and ferns; 4–6, gymnosperms; 7–30, angiosperms. Reprinted by permission.

finer stratigraphic scale between the boundary-bracketing leaf quarries served as a proxy for the megafossil flora and pinpointed the position of the K–T boundary, which was verified by the presence of an iridium anomaly. Megafloral turnover across the boundary was initially estimated to be 79%; palynological extinction at the boundary was estimated to be 30%. The difference between these numbers, both derived from plant fossils, is due to the different levels of taxonomic resolution of the kinds of plant fossils evaluated (see Section 3.3). At Pyramid Butte, the stratigraphically highest Maastrichtian pollen assemblages are present in a coal bed, and the lowermost Paleocene assemblages are present in mudstone just above the coal (Figure 6.6). This pattern is the reverse of that at most other localities in North America where the palynological extinction has been described, in which a coal bed overlies mudstone or shale at the K–T boundary. Prior to the discovery of the Pyramid Butte site, it could be speculated that the apparent replacement of an angiosperm-rich palynoflora by one dominated by ferns and sphagnum moss was due to the presence of a coal bed above the K–T boundary. At Pyramid Butte, characteristic Maastrichtian pollen is present in coal but is absent in the

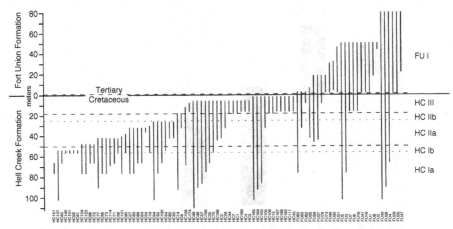

Figure 6.7 Biostratigraphic range chart for all megafloral taxa that occur at more than one stratigraphic level in the composite section in the Marmarth, North Dakota, area (from Johnson and Hickey 1990). Numbers across bottom designate leaf morphotypes; megafloral zone and subzone names are at the right (HC = Hell Creek, FU = Fort Union). Megafloral zone boundaries are indicated by dashed lines; subzone boundaries are indicated by dotted lines. Reprinted by permission.

overlying mudstone facies, just above the level at which an iridium anomaly and shocked quartz are found.

In 1990, the second of the so-called "Snowbird conference" symposium volumes was published (for the complete set of publications in this authoritative series, see Silver and Schultz 1982, Sharpton and Ward 1990, Ryder *et al.* 1996, and Koeberl and MacLeod 2002). Two papers appeared in the 1990 volume that concerned the Marmarth area. Johnson and Hickey (1990) discussed the accumulated data from 57 Cretaceous and 30 Paleocene localities that had yielded 11 503 specimens assigned to 247 morphotypes of plant megafossils. They described four megafloral assemblage zones, three in the Hell Creek Formation and one in the Fort Union Formation. The extent of megafloral change across the K–T boundary (essentially the Hell Creek–Fort Union formational contact) was presented as 79% and was based on a direct comparison of the composition of the uppermost Maastrichtian megafloral zone Hell Creek (HC) III with that of the lowermost Paleocene Fort Union (FU) I (Figure 6.7). Leo Hickey's coauthorship on this paper signified his conversion from the position of a skeptic of the K–T plant extinction story (see Section 4.5) to that of a supporter of the theory. The second paper of direct interest in the 1990 volume is that of Nichols and Fleming (1990). These authors reviewed the palynological data on the K–T boundary that had developed from 1982 to 1990 from sites that had both a palynological extinction horizon and an iridium anomaly. They

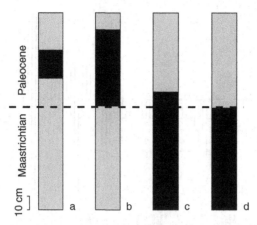

Figure 6.8 Diagrams of stratigraphic relations of clastic rocks (gray), coal beds (black), and the K–T boundary (dashed line) at different localities in western North America (modified from Nichols and Fleming 1990). a – City of Raton, New Mexico; b – Morgan Creek, Saskatchewan; c – Sugarite, New Mexico; d – Pyramid Butte, North Dakota. With changes in vertical scale, these diagrams also pertain to other localities discussed later: a, Dogie Creek, Wyoming; b, Starkville South, Colorado, and localities in the Hell Creek area, Montana; c, Police Island, Northwest Territories. Reprinted by permission.

summarized the data from the Pyramid Butte locality and compared it with patterns of extinction at the K–T boundary in three other areas in western North America (to be discussed later). Nichols and Fleming emphasized the independence of the palynofloral change at the K–T boundary from facies changes (mudstone to coal, or coal to mudstone) in their summaries of the localities; these relationships (Figure 6.8) also pertain to other localities, as will become evident in Chapter 7.

Johnson (1992) published additional information about the megaflora of the Hell Creek Formation in southwestern North Dakota. In that report, he showed that 90% of all the megafloral taxa (both Maastrichtian and Paleocene) were angiosperms. While other K–T boundary localities were being discovered in that area and detailed paleobotanical and palynological data were being gathered at them, a somewhat parallel study was undertaken in the Hell Creek and Fort Union formations farther to the east in the Williston Basin, in south-central North Dakota. Murphy *et al.* (1995) published the results of that study, which was more of a reconnaissance than were those in southwestern North Dakota. The K–T boundary was bracketed palynologically in 12 of 32 measured sections, within 30 cm at 8 of them, but closely spaced samples were not collected once the approximate position of the boundary had been determined. The K–T boundary layer with its iridium and shocked quartz was not found, and no paleobotanical collections were made. The palynoflora in south-central North Dakota is the same as that in the southwestern part of the state, and 16 familiar Cretaceous taxa were found in the Maastrichtian samples.

The most comprehensive and complete description of the K–T megaflora in southwestern North Dakota is published in the *Geological Society of America Special Paper* 361 (Johnson 2002). By 2002, the Williston Basin database included 158 localities (106 Maastrichtian and 52 Paleocene) and 13 571 specimens in 380 morphotypes or species (Figures 6.9 to 6.12). The stratigraphic context for these fossils had been delimited using palynostratigraphy, magnetostratigraphy, and vertebrate paleontology, all tied to the K–T boundary in a stratigraphic frame-work composed of 37 measured sections. Such a massive database is difficult to grasp, or even to portray graphically. Elimination of morphotypes present at only one locality or at only one or two stratigraphic levels left 98 stratigraphi-cally robust morphotypes to be plotted in 54 levels (Johnson 2002). Johnson defined five megafloral assemblage zones by subdivision of two of his three original Maastrichtian zones. He noted that extinction involved all of the numerically dominant plant taxa of the upper part of the Hell Creek Formation, and that survivorship appeared to be greatest among plants that had inhabited Maastrichtian mires. He discussed the effects of climate change (warming at the very end of the Maastrichtian) and base-level change (caused by brief re-advances of the regressing Western Interior seaway) on the terminal Cretaceous flora.

Kroeger (2002) analyzed the palynoflora of the Hell Creek Formation in northwestern South Dakota with relation to paleoenvironments. He distin-guished three kinds of flood-basin deposits (lake, marsh, and crevasse-splay) and four kinds of meander-belt deposits (channel, point-bar, point-bar swale, and abandoned channel) in the Hell Creek Formation in his study area. Kroeger identified 44 palynomorph taxa in 34 samples from these deposits, and using detrended correspondence analysis, he identified six associations of taxa within certain deposits. Some of the taxa are somewhat more common in either the flood-basin or the meander-belt facies; others tend to occur preferentially in marsh or lake deposits. Hence, some influence of facies on occurrences of certain palynofloral species is evident, although certainly not enough to account for the extinctions at the K–T boundary (as clearly indicated in Figure 6.8).

The differences in relative abundance observed by Kroeger (2002) were inter-preted as reflecting differing plant communities that characterized specific paleoenvironments. These results are highly significant because they may account for sample-to-sample variations in palynomorph content within closely spaced samples from a single measured section. Such variations are often

Figure 6.9 Representatives of the lobe-leafed Hell Creek megaflora (specimen catalogue numbers given in parentheses). a – Magnoliopsida HC 060 (DMNH 19233), b – *Araliaephyllum polevoi* (DMNH 8458), c – *Erlingdorfia montana* (DMNH 6253), d – *"Artocarpus" lessigiana* HC 179 (DMNH 13650), e – Magnoliopsida HC 199 (DMNH 13634), f – *Cissites panduratus* (DMNH 6240), g – *Bisonia niemii* (DMNH 16694). Scale bar is 1 cm.

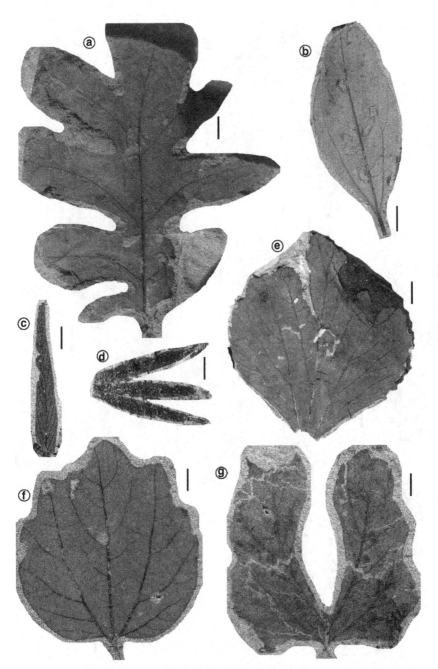

Figure 6.10 Representatives of the Hell Creek megaflora. a – *"Artocarpus" lessigiana* HC 179 (Pioneer Trails Regional Museum spec. 2559), b – *Marmarthia pearsonii* HC 162 (DMNH 7713), c – *Dryophyllum tennesseensis* HC 044 (YPM 6167), d – *Palaeoaster inquirenda* HC 007 (YPM 6399), e – *Platanites marginata* HC106 (YPM 6234), f – *Marmarthia trivialis* HC 105 (DMNH 7395), g – *Liriodendrites bradacii* HC166 (DMNH 6889). Scale bar is 1 cm.

Figure 6.11 Representatives of the Hell Creek megaflora. a – *Ginkgo adiantoides* HC 114 (DMNH 18927), b – *Nilssonia yukonensis* HC 164 (Field Museum of Natural History spec. 6295), c – *Taxodium olrikii* HC 071 (YPM 6192), d – *Fokieniopsis catenulata* HC 137 (DMNH 19428), e – *Glyptostrobus* sp. #2 HC 009 (YPM 6133), f – *Metasequoia* sp. #2 HC 035, g – *Metasequoia occidentalis* FU 003 (YPM 6053). Scale bar is 1 cm.

Figure 6.12 Representatives of the FUI megaflora from the central Great Plains.
a – *Penosphyllum cordatum* (DMNH 20033), b – *Zizyphoides flabella* (DMNH 21874),
c – *"Populus" nebrascensis* (YPM 7273), d – *Browniea serrata* (DMNH 22706), e –
Cornophyllum newberryi (YPM 6078), f – *Zizyphoides flabella* (YPM 7289), g – *Cornophyllum newberryi* (DMNH 21891), h – *Paranymphaea crassifolia* (YPM 7266). Scale bar is 1 cm.

observed but seldom explained. Their importance is great in paleoecological studies, but less so in biostratigraphy. With respect to taxa that appear to become extinct at the K–T boundary in northwestern South Dakota, Kroeger noted that some are paleoenvironmentally sensitive, but others are not. Species in the extinction palynoflora Kroeger regarded as paleoenvironmentally sensitive (i.e., facies-controlled) include *Liliacidites complexus*, *Orbiculapollis lucidus*, and *Tricolpites microreticulatus*. Species of *Aquilapollenites* evidently are independent of paleoenvironmental influence, as their occurrence is not determined by sedimentary facies.

Nichols (2002a) based his description of the Hell Creek palynoflora on 110 000 specimens from 20 measured sections; he described 98 of about 115 palynofloral taxa, a few of which are illustrated here in Figures 6.13 and 6.14. The palynoflora is numerically dominated by angiosperm pollen, although it includes significant percentages of gymnosperm pollen and spores of ferns and other cryptogams. The palynofloral taxa Nichols described include 69 angiosperms, 7 gymnosperms, 19 pteridophytes, and 3 bryophytes. Of these, 33 – about one-third of the most commonly occurring pollen species – do not occur in rocks of Paleocene age in the region. For this reason, they are called "K taxa." Five of the K taxa disappear within the Hell Creek Formation well below the K–T boundary, and their extinction cannot be attributed to the K–T boundary event. Three other K taxa are among the most commonly occurring taxa in the Hell Creek Formation. The K taxa whose disappearance marks the K–T boundary in the Williston Basin are listed in Table 6.1. The most commonly occurring taxa in the Hell Creek Formation, those present at 70% or more of the localities sampled in southwestern North Dakota, include ten taxa of angiosperm pollen, two of gymnosperm pollen, four of fern spores, and one of bryophyte spores (eight of these are genus-level taxa). The widespread occurrence of these palynomorph taxa suggests that, although angiosperms dominated the taxonomic diversity of the flora, ferns and bryophytes were also prevalent in the Hell Creek paleoecosystem, and certain gymnosperms were well represented. Nichols found that 25 of his Maastrichtian palynomorph species (K taxa) occur as much as 2.7 m above the top of the Hell Creek Formation, in a basal part of the Fort Union Formation that is Maastrichtian in age. This important observation further demonstrates that the position of the K–T boundary as defined by palynological extinctions is independent of the lithology and facies of Hell Creek and Fort Union formations.

Our joint paper on the palynology of the K–T boundary (Nichols and Johnson 2002) presents palynological and stratigraphic data from 17 measured sections that cross the boundary, including two in which its precise position is verified by geochemical and mineralogical evidence. Data from seven additional

Aquilapollenites spp. and *Wodehouseia spinata*

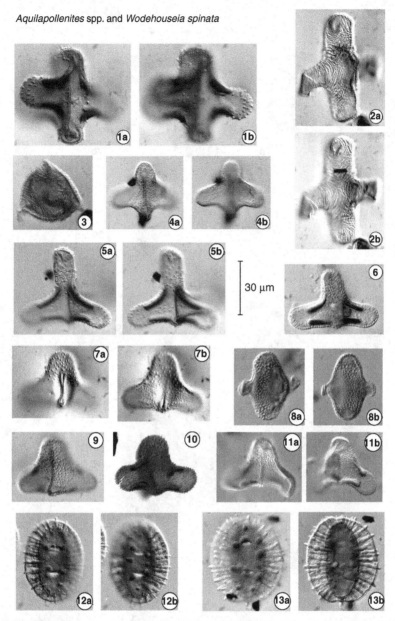

Figure 6.13 Representatives of the Hell Creek palynoflora (species of *Aquilapollenites* 1–11 and *Wodehouseia* 12–13); many specimens photographed at two levels of focus. 1a, 1b – *A. attenuatus*; 2a, 2b – *A. conatus*; 3 – *A. bertillonites*, 4a, 4b – *A. marmarthensis*; 5a, 5b – *A. collaris*; 6 – *A. delicatus*, 7a, 7b – *A. quadrilobus*; 8a, 8b – *A. reductus*; 9, 10 – *A. quadrilobus*; 11a, 11b – *A. senonicus*; 12a, 12b, 13a, 13b – *W. spinata*; symbol μm = micrometers (modified from Nichols 2002a). Reprinted by permission.

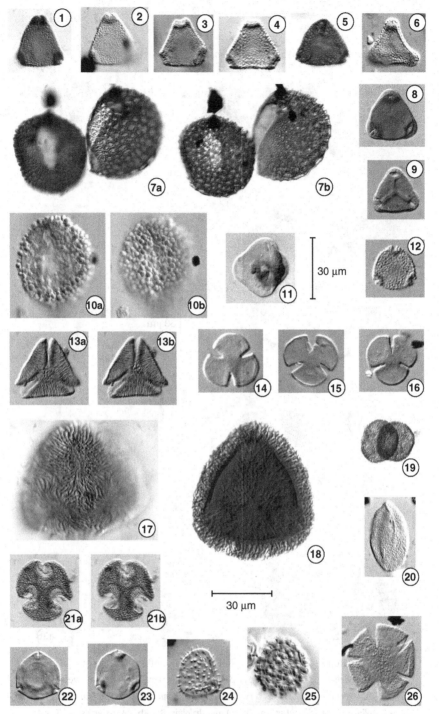

Figure 6.14 Representatives of the Hell Creek palynoflora; many specimens photographed at two levels of focus. 1–6 – *Tschudypollis* spp.; 7a, 7b – two specimens of *Liliacidites complexus*; 8 – *Myrtipites scabratus*; 9 – *Porosipollis porosus*; 10a, 10b – *Racemonocolpites formosus*; 11 – *Orbiculapollis lucidus*; 12 – *Retibrevitricolporites beccus*; 13a, 13b – *Striatellipollis striatellus*; 14–16 – *Tricolpites microreticulatus*; 17, 18 – *Styxpollenites calamitas*; 19 – *Dyadonapites reticulatus*; 20 – *Arecipites columellus*; 21a, 21b – *Libopollis jarzenii*; 22 – *Kurtzipites circularis*; 23 – *Kurtzipites trispissatus*; 24 – *Pandaniidites typicus*; 25 – *Erdtmanipollis cretaceus*; 26 – *Leptopecopites pocockii* (modified from Nichols 2002a) Reprinted by permission.

Table 6.1 *Palynomorph taxa whose extinctions mark the K–T boundary in the Williston Basin, North Dakota*

Anacolosidites rotundus Stanley 1965

Aquilapollenites attenuatus Funkhouser 1961

Aquilapollenites collaris (Tschudy and Leopold 1971) Nichols 1994

Aquilapollenites conatus Norton 1965

Aquilapollenites delicatus Stanley 1961

Aquilapollenites marmarthensis Nichols 2002

Aquilapollenites quadricretaeus Chlonova 1961

Aquilapollenites quadrilobus Rouse 1957

Aquilapollenites reductus Norton 1965

Aquilapollenites senonicus (Mtchedlishvili 1961) Tschudy and Leopold 1971

Aquilapollenites turbidus Tschudy and Leopold 1971

Aquilapollenites vulsus Sweet 1986

Cranwellia rumseyensis Srivastava 1967

Ephedripites multipartitus (Chlonova 1961) Yu, Guo and Mao 1981

Leptopecopites pocockii (Srivastava 1967) Srivastava 1978

Libopollis jarzenii Farabee *et al.* 1984

Liliacidites altimurus Leffingwell 1971

Liliacidites complexus (Stanley 1965) Leffingwell 1971

Marsypiletes cretacea Robertson 1973 emend. Robertson and Elsik 1978

Myrtipites scabratus Norton in Norton and Hall 1969

Orbiculapollis lucidus Chlonova 1961

Polyatriopollenites levis (Stanley 1965) Nichols 2002

Racemonocolpites formosus Sweet 1986

Retibrevitricolporites beccus Sweet 1986

Striatellipollis striatellus (Mtchedlishvili in Samoilovitch *et al.* 1961) Krutzsch 1969

Styxpollenites calamitas Nichols 2002

Tricolpites microreticulatus Belsky, Boltenhagen, and Potonié 1965

Tschudypollis retusus (Anderson 1960) Nichols 2002

Tschudypollis spp.

measured sections, which are near to but do not cross the boundary, supplement the data from the K–T sections, with respect to patterns of palynomorph occurrences. Our analysis is based on 700 000 specimens assigned to 110 taxa recovered from more than 350 samples. As noted above, the fundamental independence of extinction and paleoenvironment previously elucidated by Nichols and Fleming (1990) was verified within the Hell Creek–Fort Union transition. However, a "facies effect" on the palynoflora was found within a stratigraphic interval we designated as "the Fort Union strata of Maastrichtian age" (the Fort Union Formation is generally regarded as entirely Paleocene

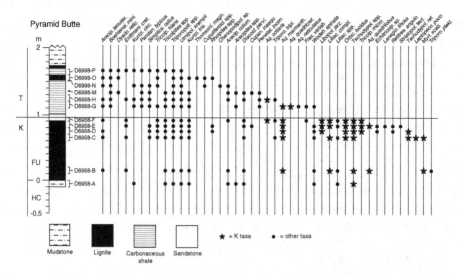

Figure 6.15 Occurrence of angiosperm pollen in samples from the Pyramid Butte section (from Nichols and Johnson 2002). HC = Hell Creek Formation, FU = Ludlow Member of the Fort Union Formation; stars designate species usually restricted to Maastrichtian rocks. Reprinted by permission.

in age). This facies effect tends to reduce the numbers and relative abundance of K taxa in lignite, carbonaceous shale, and carbonaceous mudstone below the K–T boundary, in limited areas where up to 3 m of the basal Fort Union Formation is of Late Cretaceous age. The pattern of palynomorph occurrences across the boundary in the Marmarth area reveals several significant aspects of the record of extinction and survival of Maastrichtian plants as represented by their pollen. All 24 of our localities in the southwestern North Dakota study area are described in Nichols and Johnson (2002), but only six selected examples are discussed here: Pyramid Butte, Mud Buttes, Torosaurus Section, Terry's Fort Union Dinosaur, Dean's High Dinosaur, and New Facet Boundary.

Pyramid Butte (locality 1), originally described in 1989, was restudied for the 2002 paper using new preparations of the original samples, and the data were plotted in a manner consistent with those from the other localities in the area (Figure 6.15). Note that only occurrences of angiosperm pollen are plotted in Figure 6.15. Pyramid Butte is a locality at which the precise position of the boundary is confirmed by the presence of the iridium anomaly (0.72 ppb; Johnson *et al.* 1989); shocked quartz is also present (Figure 6.16). As shown in Figure 6.15, the K–T boundary at this locality is 95 cm above the base of the Fort Union Formation, which is marked by a lignite bed 90 cm thick. The lignite bed is clearly Maastrichtian in age because it yielded a palynological assemblage that includes ten of the K taxa characteristic of the uppermost Cretaceous

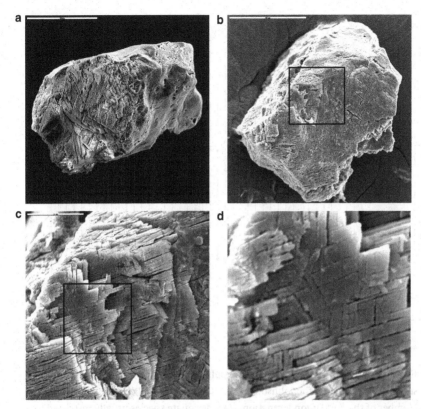

Figure 6.16 Scanning-electron micrographs of etched grains of shocked quartz from the K–T boundary at Pyramid Butte (from Nichols and Johnson 2002). Scale bars in "a" and "b" equal 50 μm. "c" is the area within the box in "b"; "d" is the area in the box in "c". Reprinted by permission.

throughout the region. Although the Pyramid Butte locality is one of the most complete K–T boundary sections in southwestern North Dakota, it does not preserve a perfect record of the boundary event. A thin bed of sandstone just above the boundary is evidence of a minor episode of erosion and redeposition just after the event. Erosion and redeposition may account for the presence of rare specimens of K taxa above the boundary at this location (Figure 6.15). The minor episode of erosion may also account for the magnitude of the iridium anomaly, which is smaller than those recorded at most other K–T boundary localities in North America, and for the absence of a fern-spore spike. Four of the K taxa present at Pyramid Butte were not observed in the uppermost lignite sample, among 2000 specimens scanned. Their absence may be due to the facies effect, that is, the plants that produced them were rare in the mire that formed the lignite. Their absence may also demonstrate the Signor–Lipps effect, discussed previously (see Section 1.3).

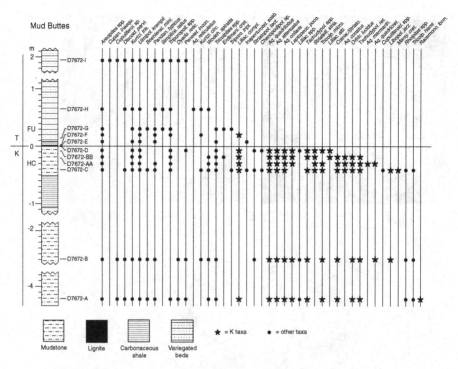

Figure 6.17 Occurrence of angiosperm pollen in samples from the Mud Buttes section (from Nichols and Johnson 2002). HC = Hell Creek Formation, FU = Ludlow Member of the Fort Union Formation; stars designate species usually restricted to Maastrichtian rocks. Reprinted by permission.

It is instructive to compare the pattern of fossil pollen occurrences at Mud Buttes (locality 5; Figure 6.17) with that at Pyramid Butte. At both localities, the precise position of the boundary is also known from the physical (geochemical and mineralogical) evidence as well as from the palynological extinction. The physical evidence of the K–T boundary at the Mud Buttes locality includes an iridium anomaly of 1.38 ppb, shocked quartz (Figure 6.18), and smectitic spherules altered from tektite glass (Figure 6.19). There is a very thin (ca. 1 cm) lignite bed just above the K–T boundary. The position of the boundary is marked by the disappearance (within 20 cm) of 14 K taxa and confirmed by the presence of impact debris. There is no evidence of a facies effect on the palynoflora at Mud Buttes, that is, there is no appreciable decline in the number or relative abundance of K taxa within the 50-cm interval of mudstone below the boundary (Figure 6.17). Note that only occurrences of angiosperm pollen are plotted in Figure 6.17. A single specimen of one K taxon was found about 2 cm above the K–T boundary; presumably, it is a reworked specimen. Figure 6.20 shows an abrupt but temporary increase in the relative abundance of fern spores just

Figure 6.18 Scanning-electron micrographs of etched grains of shocked quartz from the K–T boundary at Mud Buttes (from Nichols and Johnson 2002). Individual grains measure 110 to 190 μm. "b" is enlargement of part of "a". Reprinted by permission.

above the boundary – this is a fern-spore spike of the kind discussed in Section 5.3. The Mud Buttes locality is the most complete K–T boundary locality in the Williston Basin and the second best in North America. It has a strong (30%) palynologic extinction at the boundary and a fern-spore spike just above; a boundary claystone layer with associated iridium anomaly, shocked quartz, and spherules; plant megafossils from the richest known Cretaceous leaf quarry in the world (the single quarry yielded more than 85 species of leaves, Johnson 2002), as well as a series of Paleocene leaf quarries; vertebrate fossils including those used to argue for the highest stratigraphic occurrence of Cretaceous dinosaurs (Sheehan *et al.* 2000, Pearson *et al.* 2001); and paleomagnetic data placing it within subchron C29r (Hicks *et al.* 2002). The only feature this section lacks is a radiometrically dated horizon.

The pattern at Torosaurus Section (locality 17; Figure 6.21) is similar to that at the Mud Buttes locality, but there are minor differences in the stratigraphy of these localities. No lignite bed is present in close proximity to the K–T boundary, which is identified solely by the palynological extinction (no analyses were

Figure 6.19 Spherules of tektite origin from the boundary claystone layer at the Mud Buttes locality. a – small hand specimen; b – polished section.

conducted to detect an iridium anomaly or shocked quartz). Ten K taxa disappear within 50 cm below the K–T boundary, five of them within 5 cm. There is no evidence of a facies effect (the lithology is a homogeneous mudstone throughout the 3 m of the 25-m measured section shown in Figure 6.21), but the Signor–Lipps effect is expressed in the pattern of highest stratigraphic occurrences of the less common K taxa. This locality takes its name from the skeleton of the ceratopsian dinosaur *Torosaurus latus* that was excavated here by the Pioneer Trails Regional Museum of Bowman, North Dakota. We collected and analyzed palynological

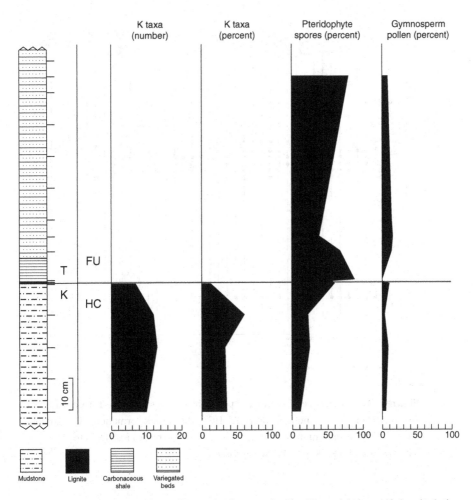

Figure 6.20 Numbers of K taxa (pollen restricted to Maastrichtian rocks) and relative abundances of K taxa, pteridophyte spores, and gymnosperm pollen in part of the Mud Buttes section (from Nichols and Johnson 2002). Strong increase in abundance of pteridophyte spores just above K–T boundary is the fern-spore "spike." Reprinted by permission.

samples to verify the latest Maastrichtian age of the dinosaur. It also proved to be an interesting K–T boundary locality because of the absence of coal or coaly facies near the boundary. Note that occurrences of spores and gymnosperm pollen as well as angiosperm pollen are plotted in Figure 6.21.

Terry's Fort Union Dinosaur (locality 10; Figure 6.22) is particularly interesting with regard to the occurrence of dinosaur remains near the K–T boundary. The boundary at this locality is 260.5 ± 2.5 cm above the Hell Creek–Fort Union contact, within strata of Fort Union lithology but latest Maastrichtian age. An associated partial skeleton of a ceratopsian dinosaur (Pioneer Trails Regional

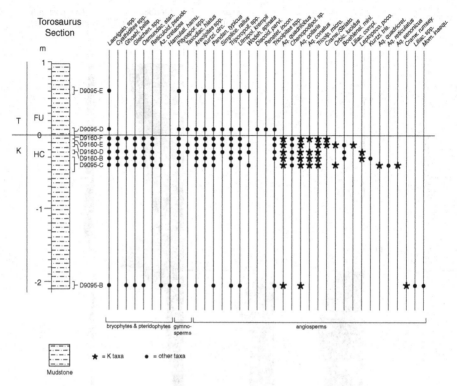

Figure 6.21 Occurrence of spores and pollen in samples from the Torosaurus Section (from Nichols and Johnson 2002). HC = Hell Creek Formation, FU = Ludlow Member of the Fort Union Formation; stars designate species restricted to Maastrichtian rocks. Reprinted by permission.

Museum specimen V96019) was collected at this locality from the Fort Union strata of Maastrichtian age, 176.5 ± 2.5 cm below the K–T boundary (Pearson *et al.* 2001; Pearson *et al.* 2002). Figure 6.23 is a photograph of the locality taken during excavation of the dinosaur bones and collection of palynological samples. This is the stratigraphically highest dinosaur skeleton in uppermost Cretaceous rocks in the Williston Basin region – significantly, it is clearly below the K–T boundary. An abundance of K taxa (Figure 6.22) verifies the latest Cretaceous age of the dinosaur. An apparent tailing-off in the presence of K taxa below that interval is attributable to the Signor–Lipps effect.

The influence of the Signor–Lipps effect on the pattern of palynomorph occurrences near and at the K–T boundary is perhaps nowhere better illustrated than at the Dean's High Dinosaur locality (Figure 6.24). D. A. Pearson of the Pioneer Trails Regional Museum collected bones of two dinosaurs from this locality; one was a *Triceratops horridus* about three meters below the K–T

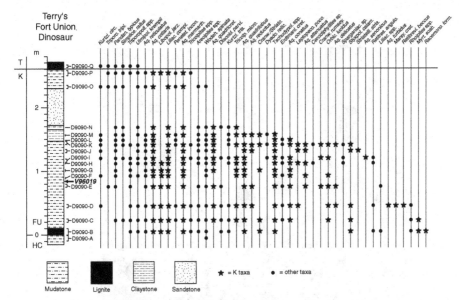

Figure 6.22 Occurrence of angiosperm pollen in samples from the Terry's Fort Union Dinosaur section, which also contained an associated partial skeleton of a ceratopsian dinosaur, Pioneer Trails Regional Museum (PTRM) specimen V96019, at arrow (from Nichols and Johnson 2002). See text for discussion. HC = Hell Creek Formation, FU = Ludlow Member of the Fort Union Formation; stars designate species restricted to Maastrichtian rocks; stippled pattern = sandstone that is barren of palynomorphs. Reprinted by permission.

boundary. The position of this dinosaur was originally believed to be closer to the boundary (hence the name of the locality) because it was only about one meter below a lignite bed in the Fort Union Formation. However, palynological analysis revealed that this lignite bed is Maastrichtian in age. The K–T boundary was sought and located at the base of the superadjacent lignite bed, about two meters higher in the stratigraphic section (Figure 6.24). The significance of Figure 6.24 with respect to the Signor–Lipps effect is that an apparent trailing off in stratigraphically highest occurrences of K taxa can be seen in both the lower and upper series of samples. Note that only occurrences of angiosperm pollen are plotted in Figure 6.24 (most spores and gymnosperm pollen range through both sampled intervals and into the Paleocene). Clearly, the lower set of samples does not define an actual extinction horizon, yet the pattern of highest stratigraphic occurrences is the same as it is about two meters above, and the same as that seen at the other K–T boundary localities discussed here (compare Figures 6.15, 6.17, 6.21, 6.22, and 6.24). Similar patterns exist for the other 12 localities described by Nichols and Johnson (2002).

Figure 6.23 Photograph of the Terry's Fort Union Dinosaur section showing the position of the associated partial skeleton of a dinosaur (PTRM V96019) within Fort Union strata of Maastrichtian age (from Nichols and Johnson 2002). The Hell Creek–Fort Union contact is at the level of the pickax head, at the base of a lignite bed; the K–T boundary is near the top of the exposure. Pioneer Trails Regional Museum paleontologist Dean A. Pearson is kneeling by the plaster jacket containing the dinosaur bones. Reprinted by permission.

New Facet Boundary (locality 16; Figure 6.25), the last of the palynologically defined K–T boundary localities in the Marmarth area to be discussed here, provides important insights into the megafloral record across the boundary in the Williston Basin. At this locality, two lignite beds near the top of a mudstone interval bracket the K–T boundary, at the top of a 2-m-thick interval of Fort Union strata of Maastrichtian age. A distinctive and unusual Cretaceous mire megaflora is present in this interval, a "Fort Union zero" (FU0) megaflora of Johnson (2002). It is more similar in composition to the typical Paleocene FUI megaflora than it is to the typical uppermost Maastrichtian HCIII megaflora of the region, but it is

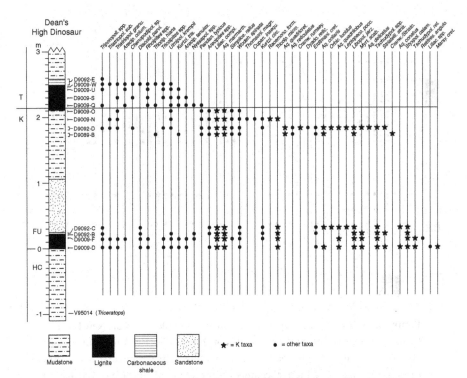

Figure 6.24 Occurrence of angiosperm pollen in samples from the Dean's High Dinosaur section (from Nichols and Johnson 2002). HC = Hell Creek Formation, FU = Ludlow Member of the Fort Union Formation; stars designate species restricted to Maastrichtian rocks. Samples were collected from two coal-bearing intervals during a search for the K–T boundary. The lower coal bed (about 1 m above a partial *Triceratops* skeleton) is Maastrichtian in age; the upper coal bed is Paleocene in age. Similar patterns of decrease in occurrences of some species below the tops of the two intervals illustrate the Signor–Lipps effect; see text for explanation. Reprinted by permission.

distinctive in floristic composition. The FU0 megaflora, which is of latest Maastrichtian age, is composed largely of plants of Paleocene aspect, along with a few species known only from the Maastrichtian. The palynological record at the New Facet Boundary locality verifies the Maastrichtian age of the FU0 megaflora. Its discovery and interpretation suggest that the Paleocene megaflora of the region is composed of Cretaceous mire vegetation that survived the K–T boundary event. Such survivor plants from Maastrichtian mires became widespread in Paleocene time, and the FUI megaflora is present in all Paleocene facies above the boundary. Note that occurrences of spores and gymnosperm pollen as well as angiosperm pollen are plotted in Figure 6.25 and that the stratigraphic position of the FU0 megaflora (DMNH 2103) is indicated.

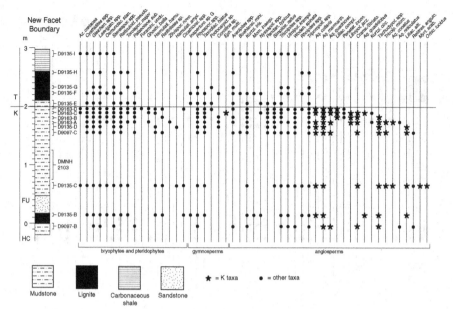

Figure 6.25 Occurrence of spores and pollen in samples from the New Facet Boundary section (from Nichols and Johnson 2002). HC = Hell Creek Formation, FU = Ludlow Member of the Fort Union Formation; stars designate species restricted to Maastrichtian rocks. Interval centered at the 1-m level and labeled DMNH 2103 yielded megafossils of Paleocene aspect preserved in a latest Maastrichtian mire paleoenvironment. Reprinted by permission.

Some of Johnson's initial interpretations of the megafossil paleobotanical record in southwestern North Dakota have changed somewhat since 2002 through the benefit of the application of quantitative methods in collaboration with paleobotanist Peter Wilf. Wilf and Johnson (2004) conducted a definitive quantitative analysis of megafloral turnover across the K–T boundary based on Johnson's data (more than 22 000 specimens from 161 localities). They found that the megaflora changed gradually during the Maastrichtian, shifted sharply at the K–T boundary, and then was static in the early Paleocene. All dominant species were lost at the boundary (verifying Johnson's 2002 conclusion), and species richness did not recover during the Paleocene. The disappearance of species more than 5 m below the boundary is attributed to normal turnover and climate change; megafloral extinction at the boundary was recalculated as 57% (cf. earlier estimates of 79–80%). The new estimate is based on only the uppermost 5 m below the boundary, and it excludes taxa that are present in only one sample. The new numbers are conservative and quite robust.

Nichols and Johnson (2002) cited the same 30% figure for palynological extinction across the boundary given in Nichols (2002a), but we have subsequently re-evaluated that number. Wilf and Johnson (2004) restricted their calculations of extinction percentage to the 5 m just below the K–T boundary. We recalculated the extinction at each of the palynological localities using this standard and found a range of 17–30%. This range is almost identical to Hotton's (2002) estimate for eastern Montana (15–30%). The plant fossil record at the K–T boundary in the eastern Montana part of the Williston Basin is discussed below.

6.3   Eastern Montana

The western part of the Williston Basin is present in eastern Montana. The strata in this area are exposed in badlands along the banks and tributaries of the Missouri River in Garfield and McCone counties and adjacent to the Yellowstone River between Miles City and Glendive. The Missouri River is dammed at Fort Peck and the Fort Peck Reservoir is surrounded by these extensive badlands (Figure 6.26). The same stratigraphic units known in western North Dakota are present in eastern Montana: the Hell Creek and Fort Union formations (Figure 6.2). In Montana, however, some authors prefer to elevate the basal Tullock Member of the Fort Union to formation rank. This region has supplied the vast majority of the vertebrate fossil record spanning the K–T

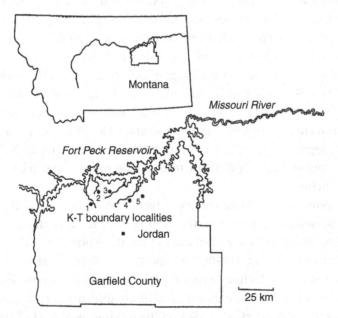

Figure 6.26 Sketch map of the Hell Creek area, Garfield County, Montana, south of the Fort Peck Reservoir (modified from Hotton 2002). Localities are: 1, Seven Blackfoot; 2, Billy Creek; 3, Herpijunk Northeast and Iridium Hill; 4, Rick's Place; 5, Lerbekmo. Reprinted by permission.

boundary and has been the site of much controversy over the timing of dinosaur extinction (Archibald 1996). Information on the megafossil paleobotanical record from eastern Montana is sparse, especially in comparison with that from western North Dakota. Palynological data are extensive, however. Some of the earliest published information on the effects of the K–T boundary event on plants anywhere in the world comes from eastern Montana, and palynology provides a well-documented picture of the extinction. A major study was conducted in the late 1980s, but was published only recently (Hotton 2002). In taxonomic detail, it exceeds the palynological study in North Dakota (Nichols 2002a), treating almost three times as many species of fossil pollen and spores, yet the changes across the K–T boundary are strikingly parallel in degree. To summarize before reviewing the key published references, a substantial portion of a diverse flora represented by angiosperm pollen suffered major decline or total extinction at the K–T boundary in this part of the Williston Basin. The boundary is marked by the well-known iridium abundance anomaly at eight localities near the Fort Peck Reservoir.

The Maastrichtian plant megafossil record in eastern Montana is relatively poorly sampled. Brown (1939) described a small flora from the Colgate Member of the Fox Hills Formation near Glendive. Shoemaker (1966) described a megaflora of 33 species from seven localities in the Hell Creek and "Tullock" (Fort Union) formations near the Fort Peck Reservoir. His data, though sparse, indicate that 85% of Hell Creek species are absent from the Tullock in his study area, and 50% are absent from the Paleocene regionally. Johnson and Hickey (1990) noted a few Maastrichtian localities in the Fort Peck region and observed that the composition of the Hell Creek and Fox Hills megafloras was similar to that observed in North Dakota. Brown (1962) recorded more than a hundred megafloral localities in Paleocene strata in eastern Montana. Because of the lack of Maastrichtian leaf localities, the megafloral record of the K–T boundary in eastern Montana is vastly under-sampled and it is difficult to conclude much about patterns of megafloral extinction.

Palynology provides the most complete record of plants across the K–T boundary in the Hell Creek area of eastern Montana (Figure 6.1), the type area of the Hell Creek Formation. The record comes from three primary investigators and four publications: Tschudy (1970), Smit and Van der Kaars (1984), Smit et al. (1987), and Hotton (2002). Tschudy's data from Seven Blackfoot Creek (locality 41) and Glendive (locality 29) published in 1970 are difficult to evaluate now because his was one of the earliest studies on the palynology of the K–T boundary in North America. He tracked the occurrences of only five individual species along with three broadly inclusive groups of pollen. The data came from a total of 29 samples from 2 localities, which encompassed up to 100 m or more of

Seven Blackfoot Creek section

Samples by locality numbers													
Hell Creek Formation									Tullock Member				
D3754-A	D3754-B	D3754-C	D3754-D	D3754-E	D3754-F	D3754-I	D3754-J	D3754-E	D3754-K	D3754-D	D3754-L	D3754-N	D3754-Q
Porosipollis porosus — range ▓													
Liliacidites complexus — range ▓													
Σ *Tschudypollis, Aquilapollenites, Tricolpites microreticulatus* — range ▓													
Leptopecopites pocockii — range ▓													
Marsypiletes cretacea — range ▓													
Wodehouseia spinata — range ▓													
Kurtzipites, Arecipites, Taxodiaceaepollenites — range ▓													
Tricolpites sp., *Triporopollenites* — range ▓													

Glendive section

Samples by locality numbers														
Hell Creek Formation													Tullock Member	
D3690-C	D3690-D	D3690-E	D3690-F	D3730-A	D3690-H	D3730-B	D3730-C	D3730-D	D3730-F	D3730-G	D3730-H	D3730-I	D3730-J	D3730-K
Porosipollis porosus — range ▓														
Liliacidites complexus — range ▓														
Σ *Tschudypollis, Aquilapollenites, Tricolpites microreticulatus* — range ▓														
Leptopecopites pocockii — range ▓														
Marsypiletes cretacea — range ▓														
Wodehouseia spinata — range ▓														
Kurtzipites, Arecipites, Taxodiaceaepollenites — range ▓														
Tricolpites sp., *Triporopollenites* — range ▓														

Figure 6.27 Stratigraphic ranges of fossil pollen species and groups in the Hell Creek Formation and Tullock Member of the Fort Union Formation at two localities in eastern Montana (modified from Tschudy 1970). Reprinted by permission.

stratigraphic section. As coarse as this analysis was, however, it clearly showed that there is a striking change in the palynoflora across the estimated position of the K–T boundary in the two measured sections (Figure 6.27). The K–T boundary was taken to be at the contact between the Hell Creek and Fort Union formations. This study demonstrated that the boundary could be identified palynologically by the disappearance of Cretaceous pollen species, which was a significant conclusion at the time. The studies by Smit and Van der Kaars (1984) and Smit et al. (1987) at Herpijunk Promontory (locality 34) established that palynological extinctions in the Hell Creek area are coincident with the K–T event as verified by the iridium anomaly. This locality also figures prominently in Hotton (2002).

Hotton's (2002) publication in *Geological Society of America Special Paper* 361 supplies the most comprehensive data from the Hell Creek area of Montana. In her dissertation, on which the 2002 paper is based, Hotton distinguished 281 palynomorph species. She studied six localities in which the position of the K–T boundary is known from the iridium anomaly. Hotton named them Seven Blackfoot (locality 40), Billy Creek (locality 36), Herpijunk Northeast (locality 35), Iridium Hill (locality 38), Rick's Place (locality 37), and Lerbekmo (locality 39); see Figure 6.26. As mentioned, localities at or near Hotton's Seven Blackfoot and Herpijunk Northeast localities also figured in the studies by Tschudy (1970) and Smit and Van der Kaars (1984), respectively. Among Hotton's taxa, 15–30% disappeared 0–2 cm above the K–T boundary, and an additional 20–30% showed significant decline in abundance. Using statistical methods, Hotton found that species diversity is unchanged from the base of the Hell Creek Formation to about 3 m below the K–T boundary. In the interval 3–4 m below the boundary, diversity and abundance of "K species" (those restricted to the Cretaceous; essentially the same group referred to as "K taxa" by Nichols) declines some-what, and the diversity of "T species" (those restricted to the Paleocene) increases. Hotton interpreted this as a facies effect due to a rise in the water table within this interval. She attributed this decline or disappearance of 30–40% of the palynoflora to the change in the depositional environment, which caused many plants that had been living in the area under conditions of lower water table to be reduced in number, or to disappear, as the water table rose. Hotton (2002) concluded that the disappearance of 15–30% of the K species was due to extinction of plants as a consequence of the impact event.

Hotton (2002) recognized roughly three times the number of palynomorph taxa in eastern Montana as did Nichols (2002a) in adjacent western North Dakota because she described all forms she encountered, regardless of how rare many of them are, whereas Nichols focused on the most commonly occur-ring forms. Significantly, their estimates of extinction at the K–T boundary are closely comparable (Nichols, North Dakota, 17–30%; Hotton, Montana, 15–30%). Both Hotton and Nichols distinguished a facies effect on the palynofloral change across the boundary in the transition from the Hell Creek to the Fort Union formations. In eastern Montana, a relatively slight change in depositional environment was noted in the uppermost part of the Hell Creek Formation, but in western North Dakota, the change was so intense that an informal stratigraphic unit – the Fort Union Formation strata of Maastrichtian age – was recognized. The root cause of the facies effect appears to be the influence of depositional environment on palynomorph assemblages described by Kroeger (2002). Hotton statistically tested the occurrence of her K species in coal vs. clastic rocks and found that about 28% of the taxa were significantly less common in or entirely absent from coal. We point out that, although lithofacies somewhat influences the occurrence of fossil pollen and spores, it does not seriously compromise their utility as indicators of geologic age, especially when they are recorded on a presence-or-absence basis. In fact, almost 65% of Hotton's K species were statistically neutral with respect to coal. Neither Hotton (2002) nor Nichols and Johnson (2002) ascribed the disappearance of characteristic Cretaceous palynomorph species to facies change. These authors concluded that the coincidence of palynofloral extinction and physical evidence of extra-terrestrial impact (the iridium abundance anomaly and shock-metamorphosed minerals in the boundary layer) are linked to a single causative event, the impact event originally hypothesized by Alvarez *et al.* (1980).

7

Other North American records

7.1 Overview

In this chapter, we discuss some of the other well-documented K–T boundary localities in North America outside of the Williston Basin. These localities are situated in other sedimentary basins in a south-to-north corridor from New Mexico to Alaska, including parts of western Canada (Figure 7.1). Collectively these localities contribute to a large database of information about plants and the K–T boundary, although it will became clear to the reader that the quality of the records varies from one to the next. The south-to-north geographic distribution of these basins constitutes a proximal-to-distal array with respect to the postulated K–T impact crater on the Yucatan Peninsula of Mexico. This fact appears to have particular relevance with respect to preservation of a boundary claystone layer and other indicators of the K–T boundary discussed in Section 2.3, such as shocked minerals.

7.2 Raton Basin, Colorado and New Mexico

The first iridium-bearing terrestrial K–T boundary sections were discovered in the Raton Basin of southeastern Colorado and northeastern New Mexico (Figure 7.2). The K–T boundary is preserved in the Raton Formation, an entirely nonmarine unit of Maastrichtian and early Paleocene age (Figure 7.3). The Raton Formation is composed of sandstone, siltstone, mudstone, coal, and minor conglomerate. Thin to thick coal beds are common in the lower and upper parts of the formation, but are scarce to absent in the middle "barren series," which is composed largely of sandstone. The K–T boundary is present in the upper part of the lower coaly interval, in some places just below sandstone of

Figure 7.1 Map of North America showing approximate locations of areas in which K–T boundary localities occur in nonmarine rocks. Each dot represents one or more (as many as 12) individual localities. Numbers are keyed to Table 2.1 and Appendix.

the barren series. It is also present in numerous locations on the eastern side of the basin, and in many places the boundary can be recognized by a distinctive, thin (1–2 cm) claystone unit visible in outcrop, even at a distance (Figure 7.4). In most places where the boundary claystone has been identified, a thin coal bed closely overlies it. An important difference in the stratigraphic setting of the K–T boundary in the Raton Basin from that in the Williston Basin is this: in the Raton Basin, the boundary is entirely enclosed within a coal-bearing interval, the Raton Formation. The fine-grained rocks of the Raton Formation contain abundant plant megafossils and palynomorphs, but with the exception of a few dinosaur tracks, vertebrate fossils are absent, likely an artifact of depositional environments unfavorable for their preservation.

In Maastrichtian time, the fluvial coastal plain in which sediments of the Raton Formation accumulated was vegetated by a diverse, angiosperm-dominated forest in which broad-leaved evergreen species were dominant and conifers were rare. The climate was warm and sub-humid. Following the K–T boundary event, the vegetation changed radically to a low-diversity forest, and

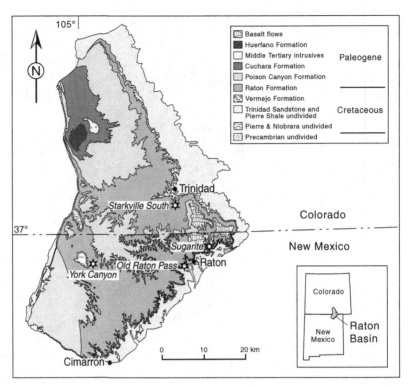

Figure 7.2 Geological map of the Raton Basin showing approximate positions of K–T boundary localities discussed in the text. In all, 13 fully documented boundary localities are known in the basin.

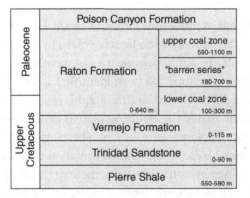

Figure 7.3 Stratigraphic nomenclature for Upper Cretaceous and Paleocene strata in the Raton Basin.

later in the early Paleocene to a somewhat more diverse angiosperm-dominated forest that was still lower in diversity than that of the latest Maastrichtian. The Paleocene climate remained warm, but became much wetter. Within a few million years after the K–T boundary, forests with rainforest physiognomy

Figure 7.4 The K–T boundary claystone (at arrow) visible even at a distance. Clear Creek North locality, Colorado.

were present. Mires in which coal-forming peat accumulated persisted from Maastrichtian to Paleocene time in the Raton Basin region.

Research on the microstratigraphic palynology of the K–T boundary began when Robert Tschudy of the US Geological Survey identified the palynological K–T boundary in a drill core as part of a search for the iridium anomaly in nonmarine rocks (Orth *et al.* 1981). Orth and his team obtained the Los Alamos–York Canyon Core (locality 42) at the York Canyon mine in New Mexico (Figure 7.2). Palynology bracketed the K–T boundary within an interval of about one meter, and gamma-ray spectrum analysis pinpointed an iridium anomaly of 5.6 ppb. Following that discovery, further palynological and nuclear geochemical (neutron activation) analyses were conducted on samples collected at 2.5-cm intervals. Results of those analyses showed that characteristic Cretaceous pollen species present low in the core abruptly disappeared precisely at the level of the peak concentration of iridium. Paleomagnetic analysis later confirmed the reversed polarity of the interval that included the K–T boundary (Shoemaker *et al.* 1987).

A short time after the discovery of a K–T boundary in the York Canyon Core, the boundary was located in outcrop exposures in the basin (Orth *et al.* 1982). The basin continued to be a prolific source of K–T boundary localities, and by 2003, about 25 had been discovered - almost all of them through the field work of Charles ("Chuck") Pillmore of the US Geological Survey. Of these, 13 have been fully documented by palynological and iridium analyses (Table 2.1).

Figure 7.5 The K–T boundary claystone (just below jackknife). A thin coal bed that lies above the claystone at this locality has been scraped away to expose the claystone layer. The jackknife is about 10 cm in length. Starkville North locality, Colorado.

The Raton sections have been the site of numerous ancillary and illustrative analyses. These include the dating of zircons from the boundary layer to show both the age of the target rock and the age of the impact event, effectively fingerprinting the Raton K–T horizon to the Chicxulub source (Kamo and Krogh 1995). Other putative K–T boundaries were identified solely by the distinctive impactite layer (Figure 7.5), and all but one of them have been verified by palynology (Table 2.1). The 13 fully documented localities are York Canyon Core (the discovery core), City of Raton (also known as Old Raton Pass), Sugarite, North Ponil, Dawson North, Crow Creek, Starkville North, Starkville South, Clear Creek North, Clear Creek South, Madrid, Berwind Canyon, and Long Canyon. The first six listed are in northeastern New Mexico; the last seven are in southeastern Colorado. For decriptions of Raton Basin outcrop localities, see Pillmore *et al.* (1984), Pillmore *et al.* (1988), Pillmore and Fleming (1990), Pillmore *et al.* (1999), and Nichols and Pillmore (2000). Selected Raton Basin localities are discussed below because they provide insights into the history of plants at the K–T boundary. The records from three localities: Starkville South, Sugarite, and City of Raton, epitomize palynological data from the K–T boundary in the Raton Basin (Figure 7.2).

Starkville South (locality 49) is the locality at which Tschudy *et al.* (1984) first found the fern-spore spike in a K–T boundary outcrop locality, shortly after it had been observed in the Los Alamos–York Canyon Core. At Starkville South, the K–T boundary is in a claystone layer just beneath a thin (5 cm) coal bed (Figure 7.6). A layer of flaky shale only millimeters in thickness at the top of

Figure 7.6 Typical K–T boundary interval in the Raton Basin showing boundary claystone layer (at tip of hammer) overlying shaly mudstone containing Maastrichtian pollen and spores, and overlain by thin coal bed and shaly mudstone containing Paleocene pollen and spores. Starkville South locality, Colorado.

the claystone layer yielded a 56 ppb iridium anomaly, the strongest ever measured in continental rocks in North America (Pillmore *et al.* 1984). The palynological extinction level is marked by the abrupt disappearance of characteristic Maastrichtian palynomorphs of what Pillmore, Tschudy, and others designated as "the *Proteacidites* assemblage." (North American species previously assigned to *Proteacidites* have been reassigned to a new genus named in honor of Robert Tschudy, *Tschudypollis*.) Members of the *Tschudypollis* (*Proteacidites*) assemblage are listed in Table 7.1. Species of the genus *Tschudypollis* are by far the most common palynomorphs in the samples below the K–T boundary. About 19% of the total Maastrichtian palynoflora disappears at the boundary. The percentage of the palynoflora that becomes extinct is low compared with that in the Williston Basin, and, in fact, the list of Raton Basin K taxa is short compared with that for North Dakota, largely because of the paucity of *Aquilapollenites* species (Table 6.1). At Starkville South, coal and mudstone just above the boundary claystone contain the fern-spore spike (Figure 7.7). Spores of a single species of the genus *Cyathidites* overwhelmingly dominate assemblages within a 10-cm interval above the K–T boundary with a peak abundance greater than 99%. This contrasts strongly with the spore content of assemblages from mudstone below the boundary, which are composed of 22–36% fern spores of several species. The boundary claystone layer itself is barren of palynomorphs.

Table 7.1 *Palynomorph taxa whose extinctions mark the K–T boundary in the Raton Basin, Colorado and New Mexico*

Aquilapollenites mtchedlishvilii Srivastava 1968 [= *A. reticulatus* (Mtchedlishvili 1961) Tschudy and Leopold 1971]

Ephedripites multipartitus (Chlonova 1961) Yu, Guo, and Mao 1981

Libopollis jarzenii Farabee *et al.* 1984

Liliacidites complexus (Stanley 1965) Leffingwell 1971

"Tilia" wodehousei Anderson 1960

Trichopeltinites sp.

Tricolpites microreticulatus Belsky, Boltenhagen, and Potonié 1965 [= "*Gunnera*"]

Trisectoris costatus Tschudy 1970

Tschudypollis retusus (Anderson 1960) Nichols 2002

Tschudypollis thalmannii (Anderson 1960) Nichols 2002

Tschudypollis spp. [= "*Proteacidites*"]

Cyathidites spores constitute 80% of the assemblage in the lower part of a 5-cm-thick coal bed overlying the boundary claystone, and 77% in the upper part. The peak abundance of fern spores (all species) is 99.5%, just above the coal. The percentage of fern spores drops close to the pre-boundary level as angiosperm pollen reappears in mudstone above the coal. This mudstone contains leaves of *Paranymphaea crassifolia*, a taxon that appears immediately above the K–T boundary from the Raton Basin all the way north to Saskatchewan.

The geologic setting of the K–T boundary at Sugarite (locality 44) is quite different from that of the other Raton Basin localities (Figure 7.2). At Sugarite, the boundary is 18 cm below the top of a 183-cm-thick coal bed (Figure 7.8). An iridium anomaly of 2.7 ppb (Pillmore *et al.* 1984) forms a double spike, indicating migration of iridium from the boundary claystone layer into the coal above and below it (Pillmore *et al.* 1999). About 17% of the characteristic Maastrichtian palynomorph taxa present in the coal below the K–T boundary disappear at the boundary. A fern-spore spike assemblage is present above the boundary and is composed of up to 78% fern spores, most of them the single species of *Cyathidites*.

The significance of the Sugarite locality for understanding the nature of the palynological record of the K–T boundary is that both the pollen extinction level and the fern-spore spike assemblage are present entirely within a coal bed. The occurrence of these phenomena at Starkville South and most other Raton Basin localities where a coal bed lies just above the boundary might suggest that certain pollen taxa disappear because of the transition from a clastic lithology to coal. The presence of coal marks a change in depositional environments inhabited by differing plant communities; an abundance of fern spores might

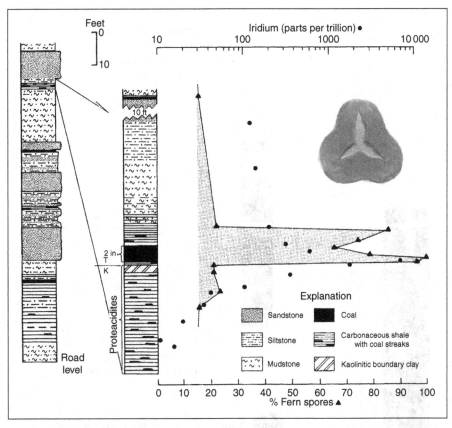

Figure 7.7 Diagram showing K-T boundary interval at the Starkville South locality with iridium concentrations (black dots) and percentages of fern spores (triangles connected by line) (modified from Pillmore *et al.* 1999). The shaded area is the "fern-spore spike," which is composed predominantly of a single species, *Cyathidites diaphana* (illustrated). Reprinted by permission.

be indicative only of a mire paleoenvironment. Sugarite disproves that interpretation and demonstrates that the pollen species that disappear at the K-T boundary were produced by plants that became extinct, and that following the extinction, the first plant communities to emerge in earliest Paleocene time were composed primarily of a low-diversity group of surviving species.

Another study of note from Sugarite is the measurement of the stable carbon isotope $\delta^{13}C$ by Beerling *et al.* (2001) that showed an appreciable 2 per mil negative excursion just above the boundary. The negative excursion was interpreted as indicative of collapse of the global carbon cycle following the K-T boundary event. A similar excursion is known from marine K-T boundary sections where it is relatively well understood. The mechanisms driving the carbon isotope excursion in this terrestrial section are not fully known, but the

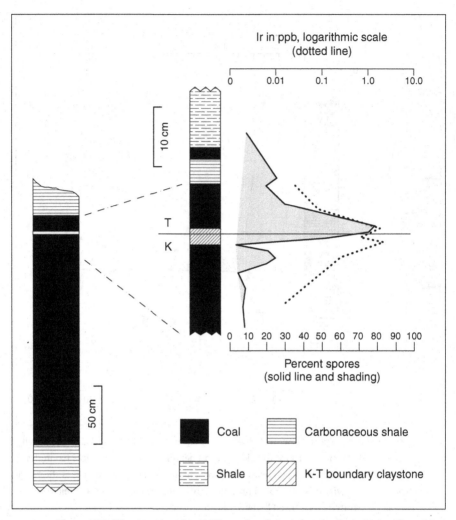

Figure 7.8 Diagram showing K–T boundary interval at the Sugarite locality with iridium concentrations (dotted line) and fern-spore spike (solid line defining shaded area) (modified from Pillmore *et al.* 1999). Note that the boundary is near the top of a coal bed about 2 m thick. Reprinted by permission.

geochemical phenomenon may well become an additional method for defining the K–T boundary in terrestrial sections.

The City of Raton site (locality 43) also demonstrates the independence of palynological occurrences from facies control at the K–T boundary. At the City of Raton locality, three thin coal beds are present within a 3.3-m-thick sequence of mudstone, siltstone, and carbonaceous shale. In succession from bottom to top, the three coal beds are 71 cm, 41 cm, and 15 cm thick. The K–T boundary is found within a claystone layer, 17 to 20 cm below the uppermost coal bed

Figure 7.9 Photographs of the City of Raton locality at a distance and close up.
a – The sign marks the locality for visitors; the iridium-bearing layer is marked by the
arrow. b – The top of the boundary claystone layer, which is several centimeters
beneath a coal bed, is at the level of the knife blade. Scale is about 15 cm long.

(Figure 7.9). Although it tends to blend in with the surrounding claystone, the
K–T boundary layer can be distinguished lithologically, and it yields an iridium
anomaly of about 1 ppb at its top (Pillmore *et al.* 1984). Thus, at this locality, the
palynological extinction level is not in close proximity to a facies change, either

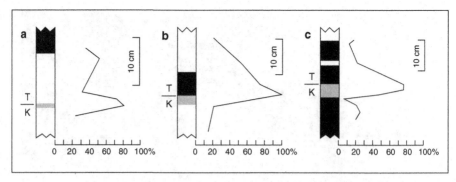

Figure 7.10 Diagrams showing the stratigraphic relationship of the fern-spore spike to the boundary claystone layer and coal beds at three localities in the Raton Basin. a – City of Raton, b – Starkville South, c – Sugarite (modified from Nichols and Fleming 1990). Reprinted by permission.

from clastics to coal or (as at Pyramid Butte in the Williston Basin) from coal to clastics. Comparison of the lithology and palynology of the Starkville South, Sugarite, and City of Raton localities shows that the palynological changes are independent of lithological changes. The relations of the palynological extinction level and fern-spore spike to lithology at these three key Raton Basin localities are shown in Figure 7.10.

Biostratigraphically important species of the Raton Basin palynoflora are illustrated in Figure 7.11. The most common species in Maastrichtian samples belong to the genus *Tschudypollis* (Figure 7.11a). Stratigraphically restricted to the Maastrichtian and present in many samples from the Raton Basin are "*Tilia*" *wodehousei* (Figure 7.11b) and *Trisectoris costatus* (Figure 7.11e). Species in common with the Maastrichtian of the Williston Basin (along with *Tschudypollis* spp.) are *Liliacidites complexus* (Figure 7.11c) and *Libopollis jarzenii* (Figure 7.11d). *Aquilapollenites mtchedlishvilii* (Figure 7.11f), which is known to many authors as *A. reticulatus*, is exceedingly rare in the Raton palynoflora, as are all species of the genus. The scarcity or absence of species of *Aquilapollenites* and the presence of species geographically more typical of the Raton Basin illustrates that the palynofloristic composition of uppermost Maastrichtian assemblages varies with latitude in western North America, a fact documented by Nichols and Sweet (1993). Figure 7.11h is the epiphyllous fungal thallus *Trichopeltinites* sp., which disappears at the K–T boundary, presumably along with the megafloral species that was its host. Also illustrated are *Cyathidites* spores (Figure 7.11g) from a fern-spore spike assemblage of earliest Paleocene age.

Megafossil paleobotanical studies have been conducted in the Raton Basin since the pioneering work of Lee and Knowlton (1917). These early studies had much influence on perceptions of the effects of the K–T impact event on plants

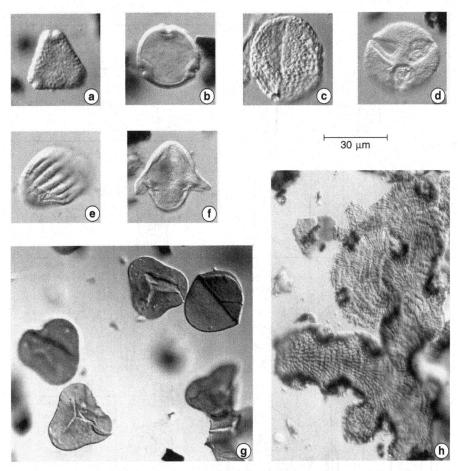

Figure 7.11 Representatives of the Raton Basin palynoflora. a – *Tschudypollis retusus*, b – *"Tilia" wodehousei*, c – *Liliacidites complexus*, d – *Libopollis jarzenii*, e – *Trisectoris costatus* (partial specimen), f – *Aquilapollenites mtchedlishvilii*, g – *Cyathidites diaphana* (several specimens in fern-spore spike), h – *Trichopeltinites* sp.

when they were reinterpreted with the addition of new data by Wolfe and Upchurch (1987a, b). Based on 48 megafloral localities in the Raton Formation, Wolfe and Upchurch (1987a) reported significant and abrupt megafloral extinction. Their collections of leaves came from 6 localities below the K–T boundary, 3 within the interval of the fern-spore spike, and 39 in the Paleocene rocks above. They divided the collections from the 39 Paleocene localities into three groups that they interpreted as representing successive phases in development of the early Paleocene flora following the K–T extinction event: angiosperm recolonization, angiosperm recovery, and an uppermost phase representing the forest that once grew in the vicinity of the York Canyon coal mine (near where the York Canyon Core had been drilled). They described five phases in changes

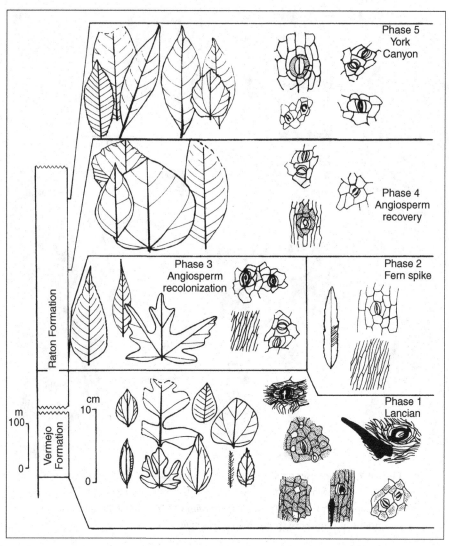

Figure 7.12 Phases in megafloral physiognomy in the Raton Basin from Maastrichtian to Paleocene time (from Wolfe and Upchurch 1987a). Reprinted by permission.

in leaf physiognomy from late Maastrichtian to early Paleocene time (Figure 7.12). Phase 1 (latest Maastrichtian) of Wolfe and Upchurch (1987a) is characterized by a megaflora of high diversity that grew in a warm, sub-humid climate. This megaflora is represented by 47 species collected from six localities, 43% of which are dicots. Phase 2 is equivalent to the fern-spore spike interval, just above the K–T boundary; it includes only two megafloral species from three localities: fern fronds and rare specimens of a bryophyte. Phase 3 is the angiosperm recolonization based on eight species from five localities, five of which are dicots. Phase 4 is called angiosperm recovery because more angiosperms appear in the megaflora, 25 species from 14 localities, 21 of which are dicots. Phase 5 includes 35 species from 20 localities in the vicinity of the York Canyon mine, 27% of which are dicots and many of which exhibit rainforest physiognomy in the form of large leaves and drip tips. Phases 3 and 4 are associated with pollen of Paleocene palynostratigraphic zones P1 and P2 (Nichols 2003). Phase 5 mega-fossils are associated with Paleocene palynostratigraphic Zone P3, which is also found associated with high-diversity megafloral rainforest sites in the Denver Basin (Johnson and Ellis 2002). This will be discussed further in the section on the Denver Basin, but the Raton Basin was the first place where possible rain-forest floras were discovered within a few million years after the K–T boundary event.

A pattern of increasing diversity in the megaflora preserved in successively higher stratigraphic levels within the Paleocene is evident from these data. However, we note that diversity increases in direct proportion to the number of localities sampled, and Wolfe and Upchurch did not report the number of specimens used in their analyses. Since diversity increases with sample size, this is a significant problem. Based on this observation, the small number of Cretaceous localities in this study probably underestimated the amount of extinction at the K–T boundary. Upchurch (personal communication 2000) reported that he and Wolfe had personally collected specimens at only two of their six Cretaceous localities, and that data for the other four were derived from analyses of the Lee and Knowlton (1917) collections, some of which included specimens from the underlying Vermejo Formation. Despite the small sample of Cretaceous megafossils, the pattern of extinction was strong. What was not clear from the data was the nature of megafloral change within the Late Cretaceous, simply because Wolfe and Upchurch (1987a) did not have adequate stratigraphic density of samples on which to base such statements.

As shown in Figure 7.12, Wolfe and Upchurch (1987a) reported summary data from analysis of dispersed leaf cuticles from palynological preparations in support of their five-phase pattern of megafloral extinction, survival, and recov-ery. While rich in promise, this cuticular work was never followed by detailed monographic work, so it is difficult to evaluate.

7.3   Denver Basin, Colorado

The Denver Basin is an asymmetrical depression adjacent to the Rocky Mountain Front Range with coarse-grained alluvial fan deposits along its wes-tern edge and fine-grained fluvial and paludal deposits extending 100 km to the east (Figure 7.13). Because of Denver's central position in the settling of the

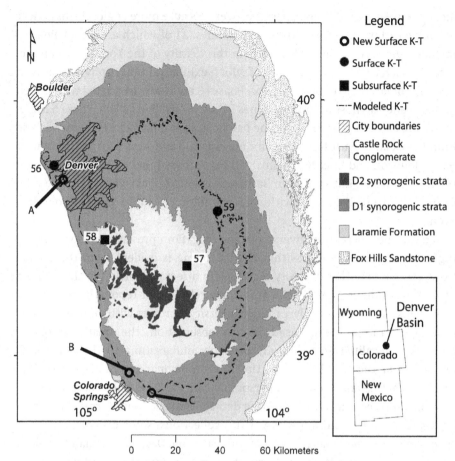

Figure 7.13 Geologic map of the Denver Basin with K – T boundary model. Shown are two cored wells (localities 57 and 58), two well-characterized K – T boundary surface outcrops (localities 56 and 59), and three poorly characterized K – T boundary surface outcrops (A – Florida and Kipling locality, B – Cottonwood Creek locality, and C – Jimmy Camp Creek locality). The modeled K – T boundary line was developed in a geographic information system (GIS) by creating a K – T boundary plane that passed through the two subsurface well points and four of the five surface points. This plane was intersected with a digital elevation model (DEM) of the Denver Basin. The dotted line on the map represents this intersection. We tested the model by sampling palynological and megafloral localities on either side of the projected K – T boundary at Cottonwood Creek (B) and found it to be correct within 70 feet at that location.

American West, Upper Cretaceous and Paleocene fossil plants have been known from the Denver Basin since the 1860s and these fossils were a major part of the Laramie Problem discussed in Section 4.2. Large compilations of these floras were provided by Knowlton (1922, 1930) and Brown (1943, 1962), but discontinuous surface outcrop made temporal sequencing of the floras difficult.

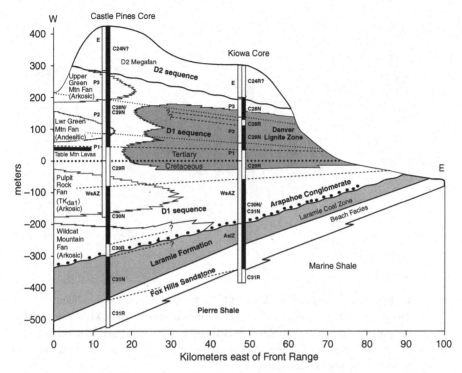

Figure 7.14 Cross section of the Denver Basin (vertical scale exaggerated) showing stratigraphic nomenclature for Maastrichtian, Paleocene, and lower Eocene strata in the Denver Basin. D1 and D2 sequence terminology after Raynolds (2002). Note that alluvial fan deposits are present on the west side of the basin while lignite deposits are present to the east. Also shown are positions of the Castle Pines and Kiowa core holes with magnetostratigraphic data from each. Palynostratigraphic zones identified in each core are indicated: E = lower Eocene zone; P1, P2, and P3 = Paleocene zones; WsAZ = *Wodehouseia spinata* Assemblage Zone (Maastrichtian); AsIZ = *Aquilapollenites striatus* Interval Zone (Maastrichtian).

The stratigraphic interval spanning the upper Maastrichtian and lower Paleocene in the basin is contained within a sequence-stratigraphic unit known as synorogenic sequence D1 (Raynolds 2002; (Figure 7.14). This unit contains both proximal and distal deposits of conglomerate, sandstone, siltstone, claystone, carbonaceous shale, and coal and is what has historically been called the Arapahoe Conglomerate, the Denver Formation, and part of the Dawson Arkose. Basaltic lava flows occur just above the K–T boundary on North and South Table mountains, near Golden, Colorado, and abundant volcanic ash beds occur in lignite seams in uppermost Cretaceous and basal Paleocene strata of the eastern part of the basin. The geochronology of this interval has been calibrated by radiometric dating of the basalt flows and ash

Figure 7.15 Photograph of the South Table Mountain locality near Golden, Colorado. The top of the butte is capped with Paleocene basalt flows. The group of people is located in a basal Paleocene channel that produces Puercan mammals and Paleocene leaves. The K–T boundary is poorly constrained at this site but is roughly 5–10 m below the lowest person.

beds (Obradovich 2002; Hicks *et al.* 2003) and by magnetostratigraphy (Hicks *et al.* 2003). Also, the basin contains a rich paleontological record from vertebrate paleontology, paleobotany, and palynology (papers cited in Johnson *et al.* 2003 and Raynolds and Johnson 2003).

The earliest resolution of the K–T boundary in the Denver Basin is that of Brown (1943, 1962). In a prescient study using the vertebrate paleontological and paleobotanical records, Brown produced a map that traced the approximate position of the boundary around the basin margin. He could not rely on a lithologic change across the boundary, as is possible to some extent in the Williston Basin, because the rocks of the K–T transition in the Denver Basin are from similar depositional environments and are extremely poorly exposed. Brown's best outcrop was on the southeast corner of South Table Mountain, near Golden, Colorado (Figure 7.15). There he collected Paleocene megaflora in association with Paleocene mammals at a site roughly 20 m above Cretaceous *Triceratops* bones. This was arguably the first terrestrial K–T boundary section known from terrestrial rocks and was significant in resolving the stratigraphic issues of the Laramie Problem.

South Table Mountain (locality 56) was also the setting for an early palynological study by Newman (1979). Because the locality is near the western edge of the basin, coarse lithologies not amenable to palynological study preclude the close spacing of samples required to determine the precise position of the K–T boundary. Newman identified characteristic Maastrichtian and Paleocene pollen in widely spaced samples and chose the approximate position of the boundary about 14 m lower than Brown had estimated its position. Kauffman *et al.* (1990) revisited the South Table Mountain locality and collected new samples for palynology and paleobotany. They narrowed the gap bracketing the K–T boundary to within a 4.9-m interval, 7–12 m below Brown's estimate and 2–7 m above Newman's. No boundary claystone layer is identifiable within this interval due to the presence of coarse sandstone.

Despite the quest having begun so early, precise placement of the K–T boundary in the Denver Basin remained elusive for almost 60 years. Since 1999, two iridium-bearing K–T boundary localities have been discovered in the Denver Basin, one in a drill core and one in outcrop. The boundary is bracketed but not pinpointed at three other localities (A, B, and C in Figure 7.13).

In 1999, under the direction of the Denver Museum of Nature and Science, a 688-m core was drilled in the axis of the basin, near the town of Kiowa, Colorado (locality 57); see Figure 7.13. The well was spudded in the Eocene part of synorogenic sequence D2 (part of the Dawson Arkose) and penetrated the entire sequence of nonmarine basin-fill to the Cretaceous marine Pierre Shale below (Raynolds and Johnson 2002); see Figure 7.14. Hundreds of fine-grained rock samples were collected for palynological and paleomagnetic analyses. The Kiowa Core and the Castle Pines Core (locality 58), which was drilled 35 km to the west in 1987, each preserve a palynostratigraphy that includes the Campanian–Maastrichtian *Aquilapollenites striatus* Interval Zone, the Maastrichtian *Wodehouseia spinata* Assemblage Zone, and lower Paleocene zones P1 to P3 (Figure 7.16), a span of about 6 Ma (Nichols and Fleming 2002).

The Laramide Orogeny was under way in the Denver Basin at K–T boundary time, which caused uplift and erosion of Cretaceous strata along the basin margin. In the early Paleocene, palynomorphs eroded from the older rocks were redeposited in the basin contemporaneously with Paleocene palynomorphs. The presence of the Cretaceous palynomorphs in Paleocene sediments constituted a significant challenge to identifying the K–T boundary using only palynology. Finally it became clear that the co-occurrence of Cretaceous marine dinoflagellates and Cretaceous species of pollen and spores could be used as evidence of reworking, and all reworked species could be disregarded. Combining this with data on first occurrences of Paleocene pollen, the palynological K–T boundary was located at a depth of 302.93 ($\pm$ 0.015) m in the Kiowa Core, within the lower part of unit D1; Figure 7.16 (Nichols and Fleming 2002).

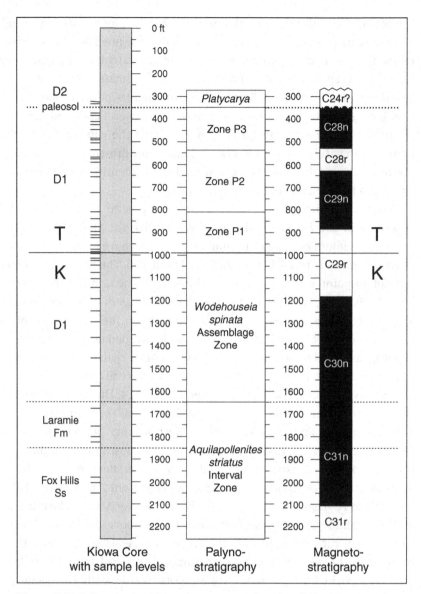

Figure 7.16 Palynostratigraphy and magnetostratigraphy of the Kiowa Core showing position of the K–T boundary at 993.85 ± 0.05 ft (302.93 ± 0.015 m) and magnetic polarity subchrons identified; depths in feet; D1 and D2 sequence terminology after Raynolds (2002).

In the Kiowa Core, the uppermost 3 mm of the Cretaceous strata yielded a diverse, angiosperm-dominated assemblage from the *Wodehouseia spinata* Assemblage Zone that includes seven characteristically Maastrichtian species not found in the 3-mm sample just above the K–T boundary (the full list of

Table 7.2 *Palynomorph taxa whose extinctions mark the K–T boundary in the Denver Basin, Colorado*

Aquilapollenites attenuatus Funkhouser 1961
Aquilapollenites collaris (Tschudy and Leopold 1971) Nichols 1994
Aquilapollenites conatus Norton 1965
Aquilapollenites delicatus Stanley 1961
Aquilapollenites mtchedlishvilii Srivastava 1968
Aquilapollenites quadrilobus Rouse 1957
Cranwellia rumseyensis Srivastava 1967
Ephedripites multipartitus (Chlonova 1961) Yu, Guo, and Mao 1981
Interpollis sp. cf. *I. supplingensis* (Pflug in Thomson and Pflug 1953) Krutzsch 1960
Libopollis jarzenii Farabee *et al.* 1984
Liliacidites altimurus Leffingwell 1971
Liliacidites complexus (Stanley 1965) Leffingwell 1971
Myrtipites scabratus Norton in Norton and Hall 1969
Retibrevitricolporites beccus Sweet 1986
"Tilia" wodehousei Anderson 1960
Tricolpites microreticulatus Belsky, Boltenhagen, and Potonié 1965
Trisectoris costatus Tschudy 1970
Tschudypollis retusus (Anderson 1960) Nichols 2002
Tschudypollis thalmannii (Anderson 1960) Nichols 2002
Tschudypollis spp.

Denver Basin K taxa is given in Table 7.2). The placement of the K–T boundary by palynology was verified by the presence of shocked quartz (Nichols 2002b). Paleomagnetic study (Hicks *et al.* 2003) showed the K–T boundary in the Kiowa Core to be just above the middle of magnetic polarity subchron C29r (Figure 7.16). A fern-spore spike is present just above the palynological extinction level (97.5% fern spores, of which more than 90% are *Cyathidites*). The abrupt change in palynofloral composition from uppermost Maastrichtian to lowermost Paleocene is striking. Assemblages in the four higher 3-mm-thick samples continue to be dominated by *Cyathidites* spores, but specimens of the fern-spore genera *Laevigatosporites* and *Toroisporis* are present in increasing numbers. A sample 9 cm above the K–T boundary appears to have a secondary fern-spore spike composed of 62% *Laevigatosporites* spores.

Once the K–T boundary had been precisely located in the Kiowa Core, its approximate position could be projected to an outcrop position on the east side of the Denver Basin using subsurface structural and stratigraphic data (Figure 7.14). Field work in one such estimated location led to the discovery of the K–T boundary at the West Bijou Site (locality 59), roughly 60 km east of the

Front Range (Barclay *et al.* 2003). We regard this locality as the most complete single K–T boundary section known in nonmarine rocks. This claim for the West Bijou Site is based on the boundary being characterized by a palynological extinction level (extinction of 21% of characteristically Maastrichtian taxa), a fern-spore spike 1 cm above the extinction level (peaking at 74% *Laevigatosporites*), an iridium anomaly (6.8 ppb), and shocked quartz (Figure 7.17). In addition to these now-familiar features, on a broader scale the locality has leaf megafloras in Maastrichtian rocks below and in Paleocene rocks above the boundary, dinosaur teeth and bones below it, and a distinctive lower Paleocene (Puercan 1) mammal jaw above it. Paleomagnetic data places it within subchron C29r, and radiometric dates from volcanic ash beds bracket the K–T boundary (Hicks *et al.* 2003).

Paleocene plant megafossils from the West Bijou locality were described in Barclay *et al.* (2003); see Figure 7.18. More than 20 Paleocene leaf localities were quarried and 4 were quantitatively collected. The dominant species at these localities are taxa that are known from the basal Paleocene sequences in New Mexico, Wyoming, Montana, the Dakotas, and Saskatchewan. Johnson and Hickey (1990) described this flora as the FUI (Fort Union I) megaflora and noted that it occurred immediately above the K–T boundary and persisted for at least 80 m into Paleocene strata. Peppe *et al.* (2005) have shown the upper limits of the FUI flora in North Dakota to be at the base of polarity subchron C28n or about 64 Ma. Interpreted as the post-fern-spike recovery megaflora, this assemblage of 20 recognizable taxa includes a group of 6 taxa that appear to dominate the vegetation at the center of the Laramide basins for a considerable period of time after the extinction event (see Table 5.1). Until 1994, this simple pattern of a rich and heterogeneous Cretaceous megaflora followed by a depauperate and widespread Paleocene megaflora seemed to be valid.

In 1994, Steve Wallace, a Colorado Department of Transportation paleontologist, discovered a significant fossil plant locality alongside Interstate 25 in Castle Rock, Colorado, on the far western margin of the Denver Basin. The Castle Rock locality is only 8 km from the mountain front and consists of a 200-m-long exposure of a buried forest floor with intact and *in situ* leaf litter and tree trunks. At first glance, the megaflora was clearly different than anything previously found in the Cretaceous and Paleogene rocks of the Rocky Mountain region. Fossil leaves from the locality were extraordinarily large (up to 61 cm long and 46 cm wide; Figure 7.19) and incredibly diverse (Figure 7.20). Excavations in 1994 and 1995 yielded more than 80 angiosperm species; many of them carried characteristic physiognomic features known from modern tropical rainforests: large leaf size, smooth leaf margins, and drip tips. At the time of discovery, the age of the locality was not known; based on

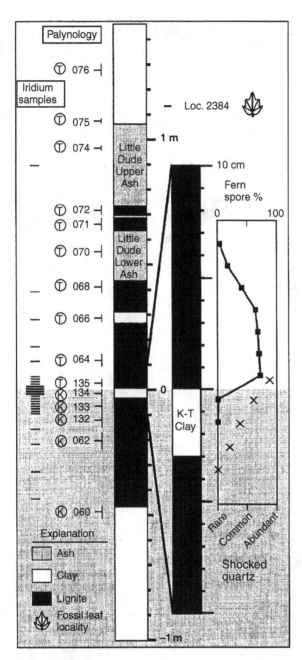

Figure 7.17 The K–T boundary interval at the West Bijou Site showing sample positions for palynology and iridium analyses, fern-spore spike as percentages of fern spores in samples, and relative abundance of shocked quartz in samples (X marks) (from Barclay *et al.* 2003). Reprinted by permission.

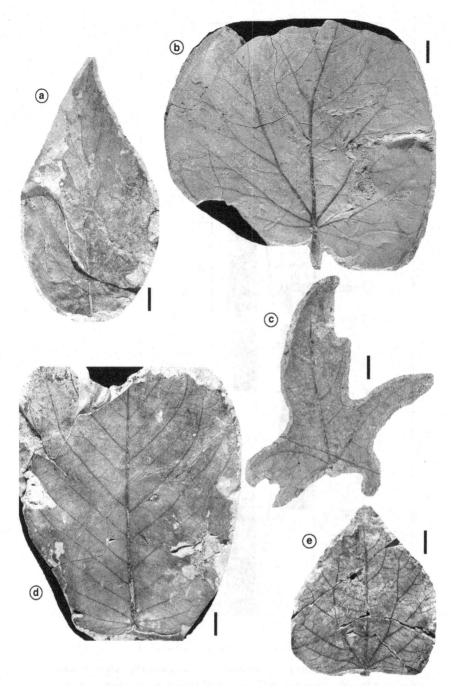

Figure 7.18 Lower Paleocene leaves from the distal part of the Denver Basin.
a – *Cornophyllum newberryi* (DMNH 23481), b – *Paranymphaea crassifolia* (DMNH 23515),
c – *Cissites panduratus* (DMNH 26144), d – *Browniea serrata* (DMNH 23502), e – *"Populus"
nebrascensis* (DMNH 23503). Scale bar is 1 cm.

Figure 7.19 Large fossil leaf (*Platanites marginata*) from the Castle Rock rainforest with Steve Wallace for scale. Steve is the vertebrate paleontologist from the Colorado Department of Transportation who originally found the Castle Rock locality.

preconceptions of the basal Paleocene FUI megaflora, knowledge of subtropical rainforest-like floras from the lower Eocene of Wyoming, and foraminiferal evidence for extreme global warming between 55 and 52 Ma, the site was misinterpreted as early Eocene in age.

Subsequent excavations revealed that the site was even more diverse and displayed spatial heterogeneity similar to extant high-diversity forests (Johnson and Ellis 2002; Ellis *et al.* 2003). Palynological analysis yielded a result of palynostratigraphic Zone P3, or lower Paleocene (Nichols and Fleming 2002). The Kiowa Core was, in part, drilled to date this enigmatic site, and results from Kiowa and the nearby Castle Pines Core in concert with additional palynology, magnetostratigraphy, radiometric dating, and basin analysis eventually led to an age determination of 63.8 ± 0.3 Ma for the site (Ellis *et al.* 2003). Several additional "rainforest" sites were discovered in proximity to the Front Range between 1994 and 2003 (Johnson *et al.* 2003) and it became clear that the early Paleocene megafloras of the western margin of the Denver Basin were considerably different from the FUI megafloras from West Bijou Creek and other localities in the central and eastern basin (compare Figure 7.21 and Figure 7.22). A preliminary basin-wide synthesis of 130 megafloral localities (Johnson *et al.* 2003) confirmed this pattern and argued for environmental, elevational, and/or orographic control for this pattern.

Figure 7.20 Line drawings to scale of the initial 98 Castle Rock leaf morphotypes from the lower Paleocene (from Johnson and Ellis 2002). The numbers next to each leaf indicate the morphotype number. Dotted lines indicate areas that are missing on the primary reference specimens but are available on other examples of the same morphotype. The upper left area contains the toothed dicot species. The upper right contains all the non-dicots (ferns, cycads, and conifers). The lower group contains the smooth-margined dicot species. Reprinted by permission.

Today, rainforests grow in areas of warm temperature, low seasonality, and high rainfall. Many extant rainforests grow on slopes of mountain ranges where orographic effects cause locally high rainfall. The proximity of the Castle Rock rainforest and associated localities to the Rocky Mountain Front Range indicates that the early Paleocene rainforests of the Denver Basin may have existed because of orographically intensified rainfall. Regional-scale climate modeling of the Rocky Mountain Front Range in the Paleogene supports this scenario (Sewall and Sloan 2006).

Figure 7.21 Paleocene leaves from Castle Rock, Colorado, proximal to the Rocky Mountain Front. a – Tiliaceae (DMNH 23246), b – Morphotype DB211 (DMNH 26021), c – Morphotype DB006 (DMNH 8698), d – Morphotype DB005 (DMNH 19180), e – Morphotype DB200 (DMNH 25513). Scale bar is 1 cm.

Figure 7.22 Paleocene leaves from the part of the Denver Basin proximal to the Rocky Mountain Front that are not from the Castle Rock locality. a – Morphotype DB909 (SMNH 102.152), b – *Rhamnus goldiana* DB018 (SMNH102.29), c – (DMNH 27553), d – *Platanus* sp. (DMNH 27554), e – Morphotype DB917 (SMNH 102.560). Scale bar is 1 cm.

Implications for the K–T boundary are less clear. If local topography and orography can have such a huge effect on the composition and diversity of early Paleocene floras, then how applicable are previous understandings of K–T boundary floral recovery based on floras collected in the middle parts of the basins? Work on Denver Basin megafloras is ongoing but several points are evident: (1) Cretaceous megafloras from the Denver Basin, although poorly sampled, are similar to typical Upper Cretaceous floras from the northern Great Plains and show little variation based on proximity to the mountains (compare Figure 7.23 and Figure 7.24); (2) the Castle Rock rainforest, although diverse, does not contain a large number of species known from Cretaceous localities; (3) the Castle Rock flora has extremely low levels of insect damage suggesting that, despite its floral diversity, the food web is still recovering from the K–T extinction (Wilf *et al.* 2006).

7.4 Powder River Basin, Wyoming

In the Powder River Basin of eastern Wyoming (Figure 7.25), the upper-most Cretaceous unit is the Lance Formation, dinosaur-bearing strata that are equivalent to the Hell Creek Formation of the Dakotas and Montana. The Lance Formation is overlain by the coal-bearing Fort Union Formation and is composed of fluvial sandstone, mudstone, and claystone. Lignite beds are rare and thin in the Lance Formation. The Fort Union Formation in Wyoming is composed of carbonaceous shale, variegated mudstone, crevasse-splay sandstone, and numerous coal beds. Some of these coal beds are of extraordinary thickness (30 m or more), especially toward the top of the formation. The strong change in the lithology of the formations at or near the K–T boundary has long been a useful criterion for approximating the position of the boundary in the field where it is assumed to be at the formation contact (Brown 1962). Three iridium-bearing K–T boundary localities are known in the Power River Basin. Two are located on the west, and one in the southeast.

The field area in the southeast part of the basin, near the town of Lusk, Wyoming, was a focus of an early study by Leffingwell (1970), who used palynology to bracket the position of the K–T boundary (locality 60). He described the morphology of characteristic species of Maastrichtian and Paleocene age and determined their relative abundances. His was one of the earliest studies to show that these palynofloras differ unmistakably and are useful in distinguishing Cretaceous from Paleogene rocks. Leffingwell found a strong change in palynofloras about 9.5 m below the top of the Lance Formation, within an interval of about 1 m bracketed by his samples. He suggested that the formational contact should be relocated to this position to coincide with the

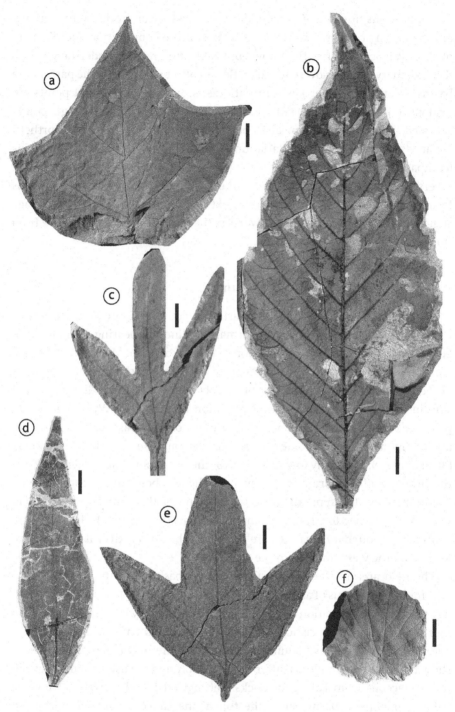

Figure 7.23 Maastrichtian leaves from the Laramie Formation and the D1 sequence in the part of the Denver Basin proximal to the Rocky Mountain Front. a – (DMNH 27557), b – Morphotype JC054 (DMNH 23731), c – (DMNH 27555), d – *Marmarthia pearsonii* (DMNH 2125), e – (DMNH 27556), f – Morphotype DB 802 (DMNH 23133). Scale bar is 1 cm.

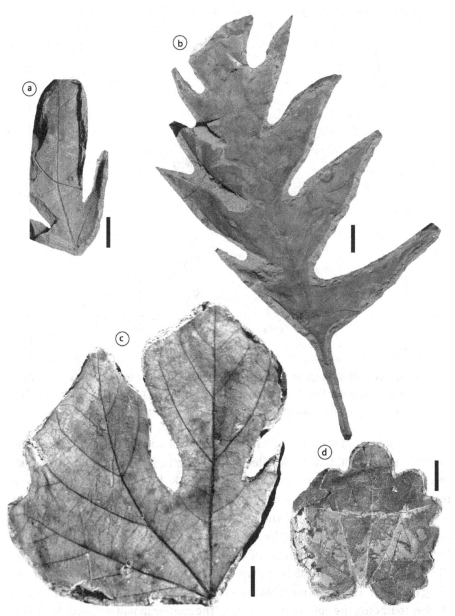

Figure 7.24 Maastrichtian leaves from the Laramie Formation and the D1 sequence from the distal part of the Denver Basin. a – Morphotype LA036 (DMNH 23150), b – "*Artocarpus*" sp. (DMNH 27558), c – (DMNH 27560), d – Marmathia johnsonii (DMNH 27559). Scale bar is 1 cm.

palynostratigraphy. Leffingwell's study was published ten years before the extraterrestrial impact theory was proposed and he postulated that the palyno-floral change was due to a depositional hiatus within the 1-m boundary interval, even though there was no field evidence for a hiatus.

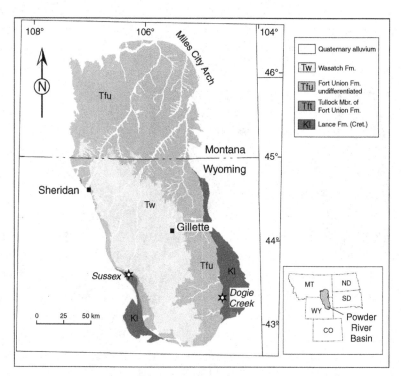

Figure 7.25 Geologic map of the Powder River Basin showing approximate positions of K–T boundary localities discussed in the text, Sussex (locality 63 in Table 2.1 and Appendix), Dogie Creek (locality 61), and Teapot Dome (locality 62) near Sussex.

No formal action was ever taken on Leffingwell's suggestion to relocate the formational contact, but his detailed locality description led Bohor and others to revisit the locality later in a search for the K–T boundary as it might be recognized by the physical evidence of impact (Bohor *et al.* 1987a). By that time, the close association of the K–T boundary with a coal deposit was well understood from the studies in the Raton Basin, the Hell Creek area, and southern Canada. That was not the setting of the boundary at this locality, however, so the first attempt to find the boundary claystone was unsuccessful. Close sampling for palynology within the interval bracketed by Leffingwell's samples was the key to finding the boundary claystone. The change in palynofloras described by Leffingwell was finally determined to center on a 3-cm-thick bed that lay 4–7 cm below a thin (3–4 cm) lignite bed (Figure 7.26). In retrospect, the stratigraphy of the boundary was more like that at the City of Raton locality in New Mexico than other localities previously discovered. The K–T boundary claystone at this locality, which was dubbed Dogie Creek (locality 61), yielded not only an iridium anomaly (21 ppb) and shocked quartz, but abundant hollow spherules, about 1 mm or less in diameter, composed of the alumino-phosphate mineral

Figure 7.26 Photographs of the Dogie Creek locality at a distance and close up.
a - Geologists are excavating the K–T boundary interval. b - The boundary claystone
is difficult to distinguish from surrounding mudstone; it lies at the level of the tip of
the hammer point and the bottom of the scale, several centimeters beneath a thin
coal bed.

goyazite (Bohor *et al.* 1987a). The goyazite was a replacement mineral; spherules
replaced by other minerals were known in marine K–T boundary localities, and
some that had been altered to kaolinite had been described at the Brownie
Butte, Montana, locality (Bohor *et al.* 1984).

Palynological analysis of the Dogie Creek locality revealed that the
Maastrichtian palynoflora disappeared within 10 cm below the base of the boundary claystone layer, half of it within 2 cm below (Figure 7.27). About 50

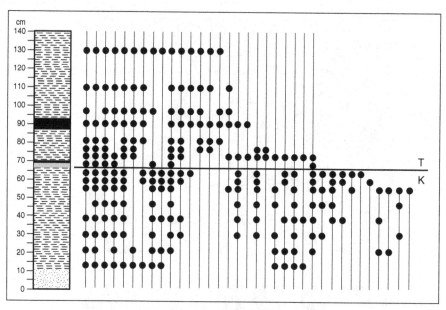

Figure 7.27 Diagram showing occurrences of most common plant microfossil taxa at the Dogie Creek locality. Gray band = boundary claystone; black bands = coal. Extensions of lines indicating stratigraphic ranges of taxa above their highest occurrence in this section are based on records from other localities.

palynomorph taxa were recorded from a suite of 17 samples, of which ten were known to be restricted to the Cretaceous. The stratigraphic relations of the extinction horizon and the boundary claystone indicate clearly that the emplacement of the boundary layer took place just after the extinction event. However, the boundary claystone contains a fern-spore spike of 93% in which a species of *Cyathidites* is dominant. The uppermost layer of the claystone layer (2–3 mm thick), which yielded iridium and shocked quartz, has a low-diversity assemblage including 63.5% fern spores. A more diverse assemblage in which fern spores are a minor component is present 4–7 cm above the K–T boundary, indicating the return of angiosperm-dominated vegetation. The presence of abundant fern spores within the boundary claystone at the Dogie Creek locality rather than above it was observed and reported by Bohor *et al.* (1987a), but no special significance was attached to it. Local transport and minor, penecontemporaneous reworking of what was considered to be the impact ejecta that formed the boundary layer were assumed to account for it, especially given the sedimentologic setting (an environment in which clastic sediment was being deposited, rather than a low-energy mire). Had the impact debris settled into a pond or mire that would later form a coal bed, little or no post-depositional transport would be expected. Such was not the situation at the Dogie Creek locality, however.

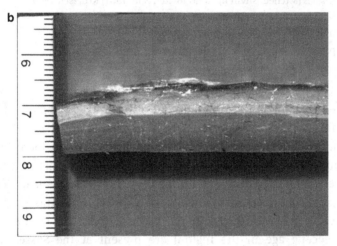

Figure 7.28 Photographs of the K–T boundary interval (a) and the boundary claystone (b) at the Sussex locality. The boundary claystone is overlain by a coal bed and underlain by mudstone. In "b" the thin, dark, uppermost layer contains the maximum iridium concentration and most abundant shocked quartz grains. The color banding of the boundary claystone visible in "b" is a local variation in its lithology not seen in "a."

Of the two K–T boundary localities on the west side of the Powder River Basin, one contributed much to the growing body of data on the effects of the K–T boundary event on plants, and the other offered a radical new interpretation that does not withstand close scrutiny. We discuss the more conventional locality first.

Nichols *et al.* (1992a) described the K–T boundary in a measured section near the abandoned town of Sussex, Wyoming (locality 63). The basal coal bed of the Fort Union is 38 cm thick at this locality. Exposure of the coal bed revealed the boundary claystone layer just beneath it (Figure 7.28). A large iridium anomaly

Table 7.3 *Palynomorph taxa whose extinctions mark the K–T boundary in the Powder River Basin, Wyoming*

Aquilapollenites attenuatus Funkhouser 1961

Aquilapollenites collaris (Tschudy and Leopold 1971) Nichols 1994

Aquilapollenites conatus Norton 1965

Aquilapollenites delicatus Stanley 1961

Aquilapollenites mtchedlishvilii Srivastava 1968

Aquilapollenites pyriformis Norton 1965

Aquilapollenites quadrilobus Rouse 1957

Cranwellia rumseyensis Srivastava 1967

Ephedripites multipartitus (Chlonova 1961) Yu, Guo and Mao 1981

Liliacidites complexus (Stanley 1965) Leffingwell 1971

Marsypiletes cretacea Robertson 1973 emend. Robertson and Elsik 1978

Striatellipollis striatellus (Mtchedlishvili in Samoilovitch *et al.* 1961) Krutzsch 1969

Tricolpites microreticulatus Belsky, Boltenhagen, and Potonié 1965

Tschudypollis retusus (Anderson 1960) Nichols 2002

Tschudypollis spp.

Wodehouseia spinata Stanley 1961

was measured (26 ppb above a background of 0.012–0.30 ppb), and shocked quartz was found. The palynologic analysis was based on 31 closely spaced samples within an interval of 1.2 m; 74 palynomorph taxa were identified. The Maastrichtian palynoflora includes most of the species known in North Dakota, but there are some differences. For example, Nichols *et al.* (1992a) identified seven species of *Aquilapollenites* at the Sussex locality; in contrast, twice that many are present in North Dakota. In all, 16 K taxa (those that do not occur in rocks of Paleocene age in the region) are present at the Sussex locality (Table 7.3), and the magnitude of extinction at the K–T boundary is 34%. The extinction is abrupt; most of the K taxa are present up to 2 cm below the boundary (Figure 7.29).

Microstratigraphic sampling of the boundary layer revealed that a few Maastrichtian specimens are present in the lowermost part of the boundary claystone layer but none in the upper part (Figure 7.28b). This suggests that the exact position of the K–T boundary can be placed within the boundary layer with millimeter precision, but such a conclusion would be over-interpretation of microstratigraphic data. The results are not reproducible from samples of the boundary claystone layer taken a few meters laterally along strike. A fern-spore spike is present at the Sussex locality. The peak of fern-spore abundance (80% of the assemblage) is in a carbonaceous layer 1–2 cm thick between the boundary claystone and the overlying coal bed; 97% of the fern spores are *Cyathidites* sp.

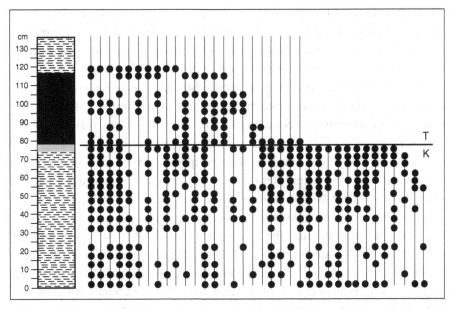

Figure 7.29 Diagram showing occurrences of most common plant microfossil taxa at the Sussex locality. Gray band = boundary claystone; black band = coal. Extensions of lines indicating stratigraphic ranges of taxa above their highest occurrence in this section are based on records from other localities.

A high percentage of fern spores is also present in the boundary claystone layer (similar to the Dogie Creek locality), but the maximum spike in abundance is 1–2 cm above. The fern-spore spike effect fades about 10 cm above the boundary claystone, but within that interval spores of two other genera of ferns, *Deltoidospora* and *Laevigatosporites*, respectively, are the most abundant palynomorphs. Evidently the fern-dominated plant communities of the earliest Paleocene coal-forming mire changed rapidly in floristic composition.

The third K–T boundary locality in the Powder River Basin, Teapot Dome (locality 62), is not far from the Sussex locality and does not differ significantly from the other two in its objectively described characteristics. However, its interpretation based on paleobotanical and palynological data (Wolfe 1991) makes it perhaps the most controversial boundary locality in North America. As at the Sussex locality, the boundary claystone is present below the basal coal bed of the Fort Union Formation. An iridium anomaly was reported, as were shocked quartz and spherules. The fossils at this locality include Cretaceous pollen in the bed below the boundary claystone layer, trilete fern spores (presumably *Cyathidites*) in a fern-spore spike, certain kinds of angiosperm pollen, and leaves and fragments of rhizomes attributed to aquatic species. Wolfe assigned the aquatic leaf species to *Nelumbites* (a Paleocene relative of lotus) and the FUI aquatic plant *Paranymphaea crassifolia*. One of the species of angiosperm pollen was identified as immature pollen of lotus (*Nelumbo*), presumably associated with the *Nelumbites* leaves, and a second was interpreted as representing a water lily such as the extant *Nuphar*. Wolfe proposed a scenario that involved two extraterrestrial impacts within four months of each other, a rain of shocked quartz grains and microtektites, an "impact winter," freezing and thawing of a lily pond, and concomitant plant extinction and survival. Wolfe specifically dated the first impact as occurring in June of the year.

Wolfe (1991) based his dating on the two species of pollen, which he interpreted as representing water lilies that had flowered and produced pollen, and lotus that was killed before producing mature pollen. In living *Nuphar* and *Nelumbo*, these events occur successively in the spring of the year. Wolfe (1991) allowed the fact that because the Paleocene ancestors of these plants may not have had exactly the same physiology, his estimate of the timing could be off by a few weeks. The problems with his hypothesis go well beyond those considerations, however. Nichols *et al.* (1992b) challenged Wolfe's scenario of events at the K–T boundary and undermined the critical element of it, the identifications of the pollen. The species Wolfe identified as pollen like that of extant *Nuphar*, the early-blooming water lily, is in fact the well-known pollen species *Pandaniidites typicus*, which has affinity not with water lilies (Nymphaeaceae), but with pandanus or screw pine (Pandancaeae) – see Hotton *et al.* (1994). Specimens Wolfe identified as tetrads of immature pollen of Paleocene *Nelumbo* bear no morphologic resemblance to modern pollen of the Nelumbonaceae. The fossils in Wolfe's samples are *Inaperturotetradites scabratus*, a species that was normally dispersed as mature, obligate tetrads. Wolfe evidently mistook specimens of *I. scabratus* for immature, unseparated tetrads. Although less common than *P. typicus*, *I. scabratus* has a well-documented record of occurrences in Upper Cretaceous and lower Paleogene rocks in the region. Specimens of these species are shown in Figure 7.30.

7.5   Wasatch Plateau, Utah

An isolated K–T boundary locality may be present at North Horn Mountain on the Wasatch Plateau of east-central Utah (locality 64). The potential of this locality for discovery of the K–T boundary is founded on its vertebrate fauna, which includes dinosaur bones, dinosaur eggs and eggshell fragments, dinosaur tracks, and mammals (Gilmore 1946, Difley and Ekdale 1999, Difley and Ekdale 2002).

The North Horn Formation, whose type locality is at North Horn Mountain, is widely distributed in the Wasatch Plateau. The K–T transition at North Horn

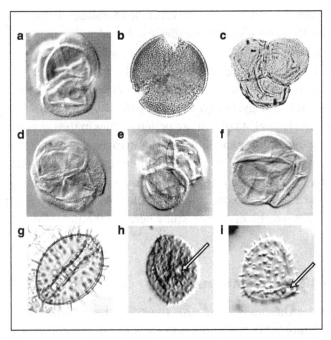

Figure 7.30 Fossil and modern pollen of disputed significance. a – inaperturate tetrad said to be immature lotus (*Nelumbo*) pollen, from a Teapot Dome sample of Wolfe (1991); b – tricolpate modern pollen of *Nelumbo lutea*, which bears no resemblance to the Teapot Dome specimens; c – holotype specimen of the inaperturate tetrad fossil pollen species *Inaperturotetradites scabratus* from the Judith River Formation (Campanian), Montana; d and e – two specimens of *I. scabratus* from the Lance Formation (Maastrichtian), Wyoming; f – specimen of *I. scabratus* from the Hell Creek Formation (Maastrichtian), North Dakota; g – monosulcate modern pollen of *Nuphar variegatum*, which bears no resemblance to specimens alleged to be fossil *Nuphar* pollen present in the Teapot Dome samples of Wolfe (1991); h – *Pandaniidites typicus* from the Lance Formation (Maastrichtian), Wyoming; i – *P. typicus* from the Hell Creek Formation (Maastrichtian), North Dakota. Note that *P. typicus*, which is the species actually present at Teapot Dome, is monoporate, not monosulcate (arrows point to the pores in "h" and "i").

Mountain is within about 400 m of mudstone with minor amounts of sandstone, algal limestone, and coal. Thin coal beds are present in the middle part of the formation, and thin limestone beds are present in the upper part. Fossils are distributed sporadically through the formation with dinosaur fossils present from the base to about the middle, where they disappear, and Paleocene mammals in the upper part (Difley and Ekdale 2002). Plant fossils include leaves and wood, which have not been studied, and palynomorphs.

Difley and Ekdale (1999) reviewed palynological data available from North Horn Mountain citing some unpublished information as well as data from their own investigations. Based on its vertebrate fauna, the lower part of the formation is Maastrichtian in age. However, assemblages from Difley and Ekdale's samples lack species of *Aquilapollenites*, *Tschudypollis*, and *Wodehouseia*, which are characteristic of Maastrichtian rocks in adjacent areas of Colorado and Wyoming. Those results agree with a study by Fouch *et al.* (1987) on the North Horn Formation where it is well exposed in Price Canyon, Utah. In their investigations of the lithology, mineralogy, and paleontology of the North Horn, Fouch *et al.* found no typical Maastrichtian pollen species in the lower part of the formation. According to Difley and Ekdale (1999), a carbonaceous interval in the middle part of the formation at North Horn Mountain contains palynomorphs indicative of a freshwater lacustrine environment (*Azolla*, *Pediastrum*), but lacks pollen definitive of either Maastrichtian or Paleocene age. The North Horn Mountain locality is worthy of further study because the palynological record is ambiguous.

7.6   Saskatchewan and Alberta

In the northernmost part of the Williston Basin in southern Saskatchewan (Figure 7.31), the uppermost Cretaceous Frenchman Formation is the lithostratigraphic and temporal equivalent of the Hell Creek Formation of Montana and North Dakota, and the overlying lower Paleocene Ravenscrag Formation is the Fort Union Formation equivalent. The base of the lowermost coal bed of the Ravenscrag Formation overlying the dinosaur-bearing Frenchman Formation approximates the position of the K–T boundary. In the valley of the Red Deer River in southern Alberta (Figure 7.31), the K–T boundary lies within the Scollard Formation, which contains coal beds more or less throughout, so the position of the K–T boundary is not so apparent, although dinosaurs such as *Tyrannosaurus* and *Triceratops* are found only in the lower part. In Coal Valley in the foothills of western Alberta, the boundary is at the base of the coal zone in the upper part of the Coalspur Formation. Stratigraphic nomenclature for the K–T boundary interval in Saskatchewan and Alberta is summarized in Figure 7.32.

Despite the fairly close geographic proximity of Saskatchewan and Alberta to Montana and North Dakota, there were noteworthy differences in the vegetation of these regions in Maastrichtian and Paleocene time. Forests composed largely of the gymnosperm family Taxodiaceae (or Cupressaceae) were far more prevalent in Saskatchewan and Alberta, and angiosperm species that produced pollen of the genus *Aquilapollenites* were more numerous. Differences in the plant communities that existed in latest Cretaceous time in Canada and the United States may account for some of the differences in interpretations of K–T boundary events in the two regions (Nichols *et al.* 1990; Nichols and Sweet 1993).

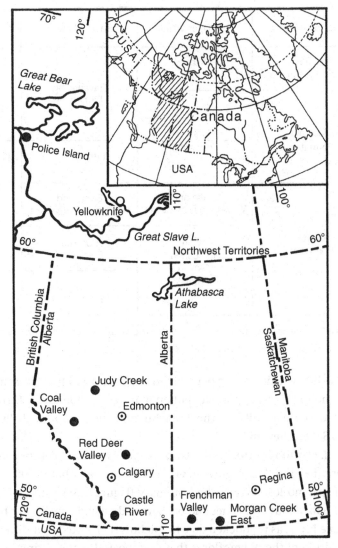

Figure 7.31 Map of western Canada showing approximate positions of K–T boundary localities (from Sweet *et al.* 1990). From north to south, localities are listed by numbers in Table 2.1 and Appendix as follows: Police Island (84), Judy Creek (67–69), Coal Valley (66), Red Deer Valley (65), Castle River (73), Frenchman Valley (75), Morgan Creek East (77). Reprinted by permission.

The earliest reports from Saskatchewan and Alberta established the fundamental nature of the K–T boundary in this region. Using palynostratigraphy in conjunction with magnetostratigraphy, Lerbekmo and Coulter (1984) determined that the K–T boundary in the Red Deer River valley in south-central Alberta (locality 65; Figure 7.33) (and the Missouri River valley in south-central

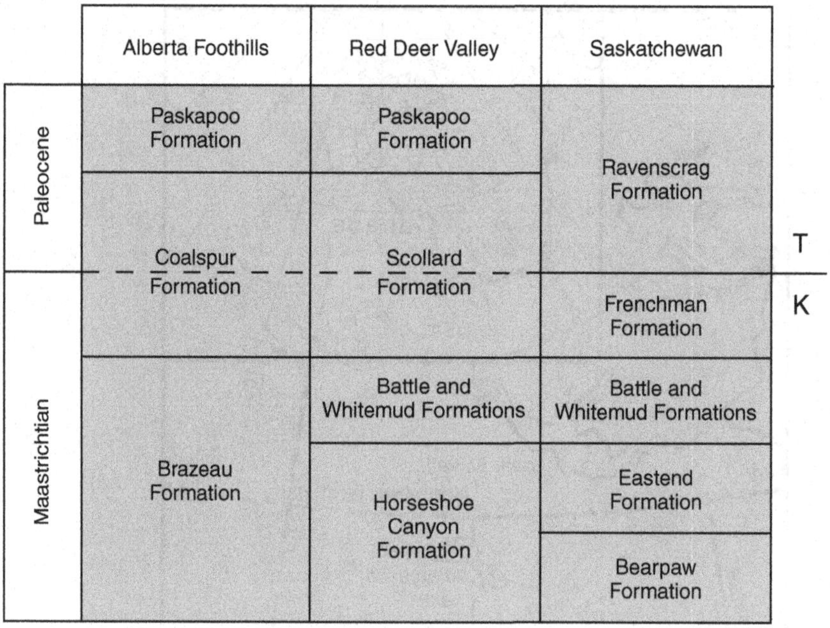

	Alberta Foothills	Red Deer Valley	Saskatchewan
Paleocene	Paskapoo Formation	Paskapoo Formation	Ravenscrag Formation
	Coalspur Formation	Scollard Formation	
Maastrichtian			Frenchman Formation
	Brazeau Formation	Battle and Whitemud Formations	Battle and Whitemud Formations
		Horseshoe Canyon Formation	Eastend Formation
			Bearpaw Formation

Figure 7.32 Stratigraphic nomenclature for the Maastrichtian and Paleocene of parts of Alberta and southern Saskatchewan.

North Dakota) lies within the upper part of magnetostratigraphic subchron C29r. The extinction of *Aquilapollenites* pollen marked the boundary. Lerbekmo (1985) repeated these results in the Frenchman Valley (Cypress Hills area; locality 75) in Saskatchewan.

Nichols *et al.* (1986) reported a K–T boundary at Morgan Creek in southern Saskatchewan (locality 76; Figure 7.34). Extinction of about 30% of the Maastrichtian palynoflora marks the boundary (Figure 7.35), and an iridium anomaly of 3 ppb, about 100 times the background level, is present at the top of the claystone layer, as is shocked quartz. Nichols *et al.* noted that differences in the composition of the palynofloras that disappeared at the extinction level in Saskatchewan and New Mexico were evidence that differing floras were simultaneously affected in much the same way by the impact event. They also noted that among the pollen taxa that survived the boundary event were two that are understood to have been produced by thermophilic (warmth-loving) plants; these were pollen of species of palms (*Arecipites*) and pandanus or screw pine (*Pandaniidites*). The persistence of this pollen across the boundary was seen as evidence that, if a period of darkness and cold resulted from an impact event as posited by Alvarez *et al.* (1980), it was of short duration, insufficient to kill off frost-sensitive plants. A fern-spore spike is present above the boundary. As reported by Nichols *et al.* (1986), the spike is composed of 96.5%

Figure 7.33 Photographs of the K–T boundary in the Scollard Formation at the Red Deer Valley locality in Alberta. Arrows point to the boundary in the distant (a) and close-up (b) views. In "b" a plastic spike has been driven into the boundary claystone layer, which is at the level of the arrow point and the hammer handle.

Laevigatosporites (but see results of subsequent studies in the vicinity of the Morgan Creek locality by Sweet and Braman 1992, discussed later). Nichols *et al.* (1986) took the presence of the fern-spore spike to confirm the work of Tschudy *et al.* (1984), who had concluded that the phenomenon was of continent-wide extent.

Beginning in 1986, a series of papers by Canadian palynologist Arthur Sweet and colleagues were published that described new localities in western Canada. Here we summarize the new data, and their interpretations that differ from those based on the studies from New Mexico, Colorado, Wyoming, Montana, and North Dakota.

Figure 7.34 Photographs of the K–T boundary at the contact between the Frenchman Formation (below) and the Ravenscrag Formation (above) at the Morgan Creek locality in Saskatchewan. Arrows point to the boundary claystone layer seen at a distance (a) and close-up (b) and the basal coal bed of the Ravenscrag is visible a few centimeters above it in "b" (scale in centimeters and inches).

Jerzykiewicz and Sweet (1986) described the K–T boundary interval in drill cores from Coal Valley in the central Alberta Foothills, at the base of the coal zone in the upper part of the Coalspur Formation (locality 66; Figure 7.36). They documented a shift in palynofloras from a diverse Maastrichtian one, in which angiosperm pollen is prominent, to lower diversity Paleocene ones, dominated by gymnosperm pollen and fern spores. They described a sequence of floral changes from latest Maastrichtian to earliest Paleocene time in Alberta and contrasted it with that documented by Tschudy *et al.* (1984) in New Mexico, Colorado, and Montana. In both Canada and the United States, a diverse latest Cretaceous angiosperm

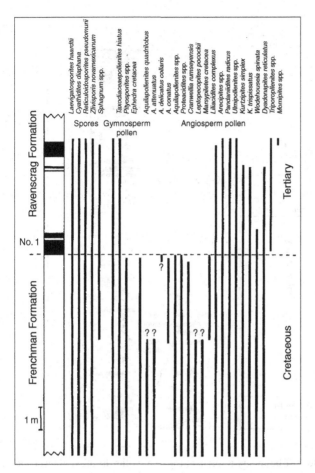

Figure 7.35 Diagram showing stratigraphic ranges of most common plant micro-fossil taxa at the Morgan Creek locality. Black bands = coal; question marks indicate uncertain limits in some stratigraphic ranges because of coarse sampling interval (from Nichols *et al.* 1986). Reprinted by permission.

flora underwent an extinction event. The species involved in the extinction varied at different localities, but notable among those disappearing were pollen types such as *Aquilapollenites* that Jerzykiewicz and Sweet (1986) characterized as morphologically exotic. Shortly thereafter, Sweet and Jerzykiewicz (1987) introduced the hypothesis that relatively large and morphologically complex pollen was produced by entomophilous (insect-pollinated) plants.

The year 1990 saw the publication of the second of the "Snowbird Conference" volumes. The new perspective based on Canadian localities was included in a paper by Sweet *et al.* (1990). They concluded that the composition of palynofloras across the K–T boundary at localities in western Canada (Figure 7.31) was influenced by sedimentary facies controlled by differences in

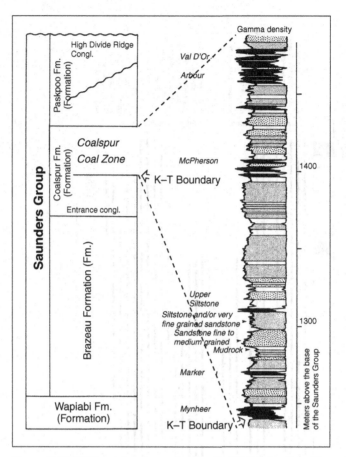

Figure 7.36 K-T boundary interval at the Coal Valley locality, Alberta Foothills (modified from Jerzykiewicz and Sweet 1986). Reprinted by permission.

precipitation and by ecological successions in plant communities that were independent of the K-T boundary event. The localities discussed in detail by Sweet *et al.* (1990) were the Judy Creek 83–313 A corehole (locality 67) and the Police Island outcrop section (locality 84). Sweet and his colleagues asserted that circumstances and events affecting plants leading up to and following the K-T boundary event were, in essence, overwhelmingly complex. The extinction event at the boundary was recognized, but it was not viewed as anything other than one in a series of complex sedimentary, paleoenvironmental, and paleoclimatic changes that occurred from latest Cretaceous to early Paleocene time.

Two years later, Sweet and Braman (1992) added observations from previously undescribed localities in western Canada (including localities 68–71, 77–79, and 82). Although they mentioned the palynofloral extinction at the boundary, their primary emphasis was on paleoenvironmental influences on palynological assemblages, not on the extinction event. Sweet and Braman documented occurrences and relative abundances of some individual species and several broader groups of palynomorphs (pollen, spores, and algal cysts) through the stratigraphic intervals at all 12 of their localities. They proposed a sweeping revision of the interpretation of the K–T boundary event and its effects on plants. Their new scenario centered on emplacement of the boundary clay-stone layer, which they interpreted to have preceded the extinction and to have consisted of more than one depositional event. From this they concluded that multiple impacts might have occurred at or around the K–T boundary.

In their 1992 paper (and subsequently), Sweet and Braman contended that palynological extinctions were not abrupt but gradual or even stepwise in pattern, approaching the K–T boundary. They stated that within the last tens of centimeters below the K–T boundary, individual samples usually yield fewer specimens and fewer species of pollen typical of the late Maastrichtian than do underlying samples. Despite this, Sweet and Braman noted that if species records from several Canadian localities are summed, most of these late Maastrichtian species are found to range up to or into the boundary claystone. That observation would appear to reinforce the concept of a significant extinc-tion at the K–T boundary. That observation is the basis for Table 7.4.

In the late 1990s, Sweet and his colleagues were involved in a major endeavor to gather lithostratigraphic, palynostratigraphic, magnetostratigraphic, and geo-chemical data relevant to events at the end of the Maastrichtian and begin-ning of the Paleocene. This was the Canadian Continental Drilling Program Cretaceous–Tertiary (K–T) Boundary Project. In this effort, three long cores were drilled in southern Canada: the Elkwater core in southeastern Alberta, the Wood Mountain core in south-central Saskatchewan, and the Turtle Mountain core in southwestern Manitoba. Sweet and colleagues presented previous research on these cores and some adjacent outcrop localities in a series of nine papers in a special issue of the *Canadian Journal of Earth Sciences* (*CJES*). Three of them put forward new data concerning the plant fossil record; they are reviewed here.

In one of the three pertinent papers in the *CJES* special issue, Braman and Sweet (1999) outlined the palynostratigraphy of the three cores. They described nine biozones or subzones from the lower Maastrichtian to the middle Paleocene, which they recognized in all three cores. The definitions of most of these palynostratigraphic biozones or subzones were compiled from previously published work, but one zone and two subzones were based on new data from the cores. The position of the K–T boundary – arguably the most significant palynostratigraphic datum in North America – is obscure in Braman and Sweet's stratigraphic scheme. It lies without emphasis between two subzones within a previously described palynostratigraphic zone (Figure 7.37).

Table 7.4 *Palynomorph taxa whose extinctions mark the K–T boundary in western Canada*

Aquilapollenites amicus Srivastava 1968

Aquilapollenites augustus Srivastava 1969

Aquilapollenites attenuatus Funkhouser 1961

Aquilapollenites collaris (Tschudy and Leopold 1971) Nichols 1994

Aquilapollenites conatus Norton 1965

Aquilapollenites oblatus Srivastava 1968

Aquilapollenites quadricretaeus Chlonova 1961

Aquilapollenites quadrilobus Rouse 1957

Aquilapollenites reductus Norton 1965

Aquilapollenites sentus Srivastava 1969

Aquilapollenites vulsus Sweet 1986

Bratzevaea amurensis (Bratzeva) Takahashi 1982

Cranwellia rumseyensis Srivastava 1967

Echiperiporites sp.

Ephedripites multipartitus (Chlonova 1961) Yu, Guo and Mao 1981

Equisetosporites lajwantis Srivastava 1968

Gabonisporis cristata (Stanley 1965) Sweet 1986

Leptopecopites pocockii (Srivastava 1967) Srivastava 1978

Liliacidites complexus (Stanley 1965) Leffingwell 1971

Marsypiletes cretacea Robertson 1973 emend. Robertson and Elsik 1978

Myrtipites scabratus Norton in Norton and Hall 1969

Orbiculapollis lucidus Chlonova 1961

Proteacidites globisporus Samoilovich 1961

Racemonocolpites formosus Sweet 1986

Retibrevitricolporites beccus Sweet 1986

Senipites drumhellerensis Srivastava 1969

Siberiapollis occulata (Samoilovich 1961) Tschudy 1971

Striatellipollis radiata Sweet 1986

Striatellipollis striatellus (Mtchedlishvili in Samoilovitch *et al.* 1961) Krutzsch 1969

Tricolpites microreticulatus Belsky, Boltenhagen, and Potonié 1965

Wodehouseia octospina Wiggins 1976

Wodehouseia quadrispina Wiggins 1976

In a second paper in the *CJES* special issue, Sweet *et al.* (1999) summarized data on relative abundances of pollen and spores from boundary localities in western Canada. They reached new conclusions about post-boundary fern-spore spikes, survival floras, and recovery floras. Adhering to views first expressed by Sweet *et al.* (1990), Sweet *et al.* (1999) concluded that the shifts in composition of pre-boundary, boundary, and post-boundary palynofloras and multi-layered boundary claystone

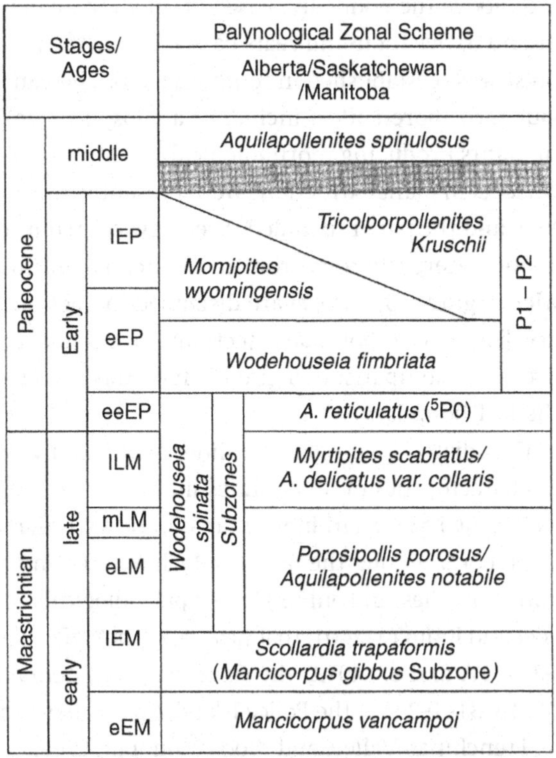

Stages/ Ages		Palynological Zonal Scheme		
		Alberta/Saskatchewan /Manitoba		

Figure 7.37 Diagram showing palynostratigraphic subdivisions of the Maastrichtian (M) and Paleocene (P) in Alberta, Saskatchewan, and Manitoba; e,E – early, early (Early); 1,L – late, late (Late); P1, P2 – palynomorph biozones (from Braman and Sweet 1999). The palynological K–T boundary, perhaps the most striking biostratigraphic horizon in western North America, is equivalent to a subzone boundary in this classification. Reprinted by permission.

deposits in western Canada are only broadly compatible with the impact theory, and that they are contrary to the catastrophic scenarios originally advocated by Tschudy *et al.* (1984) and later by Nichols and his colleagues (Nichols *et al.* 1986, Nichols *et al.* 1992a, Nichols and Fleming 1990, Fleming and Nichols 1990).

The third paper in the *CJES* special issue that is pertinent to the record of plants at the K–T boundary is perhaps the most interesting one. McIver (1999) used coal petrographic techniques to study plant fragments, especially leaf cuticle, in samples of lignite that spanned the boundary in the Wood Mountain core. Her analysis of changes in the composition and relative abundance of these paleobotanical fossils demonstrated that a pre-boundary, conifer-dominated, swamp forest was abruptly replaced by an angiosperm-dominated, herbaceous wetland, after the extinction event. She suggested that one or more of the killing agents were freezing temperatures, acid rain, thermal pulse, and shock waves. McIver saw no evidence of wildfires at or above the boundary. Based on modern analogues, she estimated that one thousand to five thousand years passed before the pre-existing taxodiaceous conifer forest was reestablished in southern Saskatchewan.

The same ideas about fern-spore spikes including angiosperm pollen, the entomophilous habit of plants producing morphologically complex pollen, and the speculative role of insects in plant extinctions were presented once again by Sweet and Braman (2001), and the paper included three new contributions. The first was detailed data and interpretations from the Police Island locality in the Northwest Territories (Figure 7.31); these are discussed in Section 7.7. The second was a summary list of K–T boundary sections in Canada and their characteristics. The third was a comparison of graphic methods used to portray palynological data at the K–T boundary.

The summary table of Canadian K–T boundary localities presented by Sweet and Braman (2001) itemized characteristics of 39 localities including the presence or absence of a boundary claystone layer, an iridium anomaly, and a stratigraphically continuous palynological record. Among the 39, only 17 have a boundary claystone, only 16 have iridium anomalies, and only 13 have a palynological record. The 13 sections listed by Sweet and Braman (2001) that have a palynological record are Judy Creek coreholes 83-313 A and 83-401A; the Canadian Continental Drilling Program (CCDP) corehole 13-31-1-2 W3; the Police Island, Coal Valley, Knudsen's Farm, Knudsen's Coulee, Frenchman Valley, and Wood Mountain Creek localities; and 4 of 6 localities in the Rock (Morgan) Creek area. These and other Canadian localities are summarized in the Appendix and evaluated in Table 2.1. The K–T "boundary" within 14 of the other sections listed by Sweet and Braman (2001) is actually not a true boundary, but a discontinuity at which Paleocene rocks sit on Maastrichtian rocks. In eight of these sections the lowermost Paleocene is missing, and at six of them part of the uppermost Maastrichtian is also missing.

The third noteworthy inclusion in the Sweet and Braman (2001) paper is a diagram comparing ways in which palynological data from North American K–T boundary localities have been published (Figure 7.38). These authors, being advocates of complexity at and near the boundary, plotted their sample-by-sample abundance data in histograms (Figure 7.38c). However, they failed to note that, unless their sample intervals were thoroughly homogenized and subsampled before splits were prepared in the laboratory (an unconventional technique in palynology), their histogram blocks unrealistically indicate uniform abundance within the entire sample interval. Abrupt changes in the widths of histogram blocks as in Figure 7.38c are not true representations of variations in relative abundance of palynomorphs from stratigraphic level to stratigraphic level, because such variations are always within gradational continua. Sweet and Braman (2001) dismissed the use of vertical lines to represent occurrences of taxa within a section

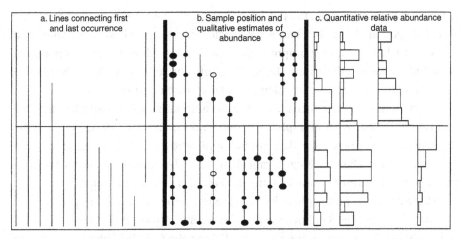

Figure 7.38 Diagram depicting alternative methods of plotting biostratigraphic data (from Sweet and Braman 2001). Reprinted with permission.

(Figure 7.38a), despite the fact that range-through data, which are of fundamental importance in all fields of biostratigraphy, are traditionally shown as vertical range lines. Enhancement of range lines by dots at sample levels where specimens were found (Figure 7.38b) is the clearest way to depict meaningful occurrence data. Given the influence of sedimentary facies in varied depositional environments and the differing potential of varied lithologies within a sampled section for preservation of palynological fossils, the relative abundance of fossils may not be as important in palynostratigraphy as the fact of their presence or absence.

Essentially no work has been done on the megafloral transition across the K–T boundary in Alberta or Saskatchewan, except that McIver and Basinger (1993) described a basal Ravenscrag Formation megaflora from Ravenscrag Butte, which contains the K–T boundary. This megaflora was similar to typical FUI megafloras from Montana, Wyoming, the Dakotas, and Colorado (Barclay et al. 2003). Subsequently, McIver (2002) described a small Late Cretaceous megaflora from the excavation site of a *Tyrannosaurus rex* near the town of Eastend, Saskatchewan. This flora contained several species in common with the typical Hell Creek megaflora described by Johnson (2002).

7.7   The Northwest Territories, Alaska, and the High Arctic

Aside from the Police Island locality in the Northwest Territories (Sweet et al. 1990, Sweet and Braman 2001), palynological and paleobotanical data from the Arctic and sub-Arctic regions of North America are meager and not based on the detailed, closely spaced samples or extensive collections of specimens required for a meaningful analysis. Some megafossil paleobotanical studies from the High

Arctic such as Spicer *et al.* (1994), Hickey *et al.* (1983), and Koch (1964) suggest that there is potential for K-T boundary studies in Alaska, Ellesmere Island, and Greenland, but much stratigraphic work remains to be done before this potential is realized. We briefly review the most significant reports from the Northwest Territories and Alaska in chronological order by date of publication.

Doerenkamp *et al.* (1976) made an effort to summarize palynological occurrence data and establish biostratigraphic zones on Banks Island and adjacent areas in the Northwest Territories. Their data are not sufficiently detailed for use in modern K-T boundary studies, but a significant change in palynofloras is indicated between their Zone CVII (upper Maastrichtian) and Zone TIa (lower Paleocene).

There are three publications by N. O. Frederiksen and colleagues pertinent to the palynology of the K-T boundary in Alaska. Frederiksen *et al.* (1988) studied 20 widely spaced outcrop samples collected along the Colville River on the North Slope of Alaska. The position of the K-T boundary was approximated, and the presence or absence of a possible unconformity at that position could not be determined. Cretaceous assemblages were dated as mid Maastrichtian, and overlying assemblages were dated as undifferentiated Paleocene. Thus, the data are not detailed enough to reveal accurately the nature of palynofloral change at the boundary. Frederiksen *et al.* (1988) suggested that many of the Maastrichtian species were insect-pollinated, as had Sweet and Jerzykiewicz (1987), a year earlier. Frederiksen *et al.* (1988) suggested that differences in Maastrichtian and Paleocene assemblages were due in part to climate change, but may also have been caused by a bolide impact.

Frederiksen (1989) discussed Maastrichtian palynofloral taxa from the Colville River samples and others collected from seismic shot holes in the vicinity. From these rather crude data, he concluded that diversity declined toward the end of the Maastrichtian under the influence of climatic deterioration. He interpreted that extinctions of Maastrichtian pollen taxa took place in a stepwise rather than gradual fashion. Frederiksen (1989) did not discuss the possibility that the apparent stepwise changes in the palynoflora were actually due to discontinuities in the sedimentary record or the wide stratigraphic separation of the samples.

Frederiksen *et al.* (1998) studied the palynoflora in a core from an oil well in the Kuparuk River Unit, the Ugnu Single Well Production Test (SWPT) on the North Slope of Alaska (locality 85 in Table 2.1 and Appendix). They regarded their data as more reliable than those published by Frederiksen *et al.* (1988) because their new samples were relatively closely spaced across the K-T boundary; these samples were collected at 15-cm intervals. (These are still widely spaced samples compared to the previously discussed core studies in the Denver Basin and southern Canada.) Occurrences of 47 taxa were tracked in 30 samples (Figure 7.39). They placed the K-T boundary where the per-sample

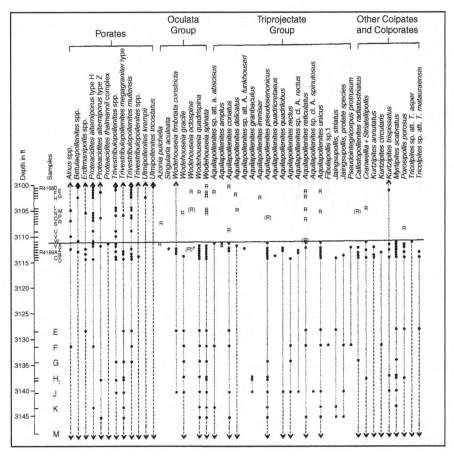

Figure 7.39 Palynostratigraphy of the K–T boundary interval on the North Slope of Alaska (from Frederiksen *et al.* 1998). Horizontal line marks position of the K–T boundary. Reprinted by permission.

diversity of Maastrichtian taxa (including some interpreted as reworked) dropped to a fraction of Cretaceous values, and where there is also a distinct drop in the ratio of strictly Maastrichtian specimens to specimens of species most common in the Paleocene. The location of the boundary was uncertain because it was bracketed within a sandy interval that may include a small disconformity. Despite rapid turnover of pollen taxa in the upper Maastrichtian (which Frederiksen *et al.* 1998 attributed to climatic deterioration), taxonomic diversity is constant until 15–30 m below the chosen boundary level. In the last 15–30 m below the boundary, last appearances of taxa are more numerous than first appearances. Nonetheless, most Maastrichtian taxa range to the top of the Cretaceous or nearly so (Figure 7.39). Frederiksen *et al.* (1998) discussed the influence of the Signor–Lipps effect (see Section 1.3) on the pattern of diversity in the uppermost Cretaceous and concluded

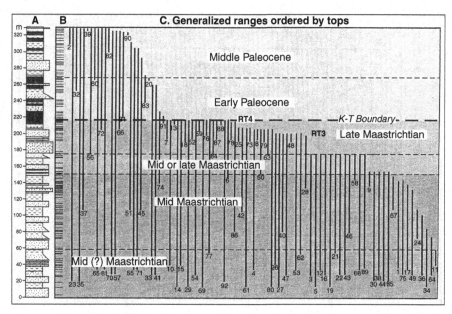

Figure 7.40 Palynostratigraphy of the K–T boundary interval at the Police Island locality, Northwest Territories (from Sweet and Braman 2001). Reprinted by permission.

that extinctions probably were more abrupt than the decline in diversity they observed seems to indicate. There was no further discussion of the supposed stepwise pattern of extinctions proposed by Frederiksen (1989). Frederiksen *et al.* (1998) compared their results from northern Alaska with those of Sweet and colleagues from the Police Island locality in the Northwest Territories. They observed that, in both sections, extinctions of Maastrichtian taxa (K taxa in our usage) take place within the uppermost meter or so of the Cretaceous. They noted that an apparent decline in diversity approaching the K–T boundary at both localities could well be attributed to the Signor–Lipps effect (see Section 1.3).

Police Island (locality 84) is the best-documented K–T boundary locality in the Arctic or sub-Arctic regions of North America (Sweet *et al.* 1990, Sweet and Braman 2001). At Police Island, the K–T boundary is marked by a palynological extinction level and verified by an iridium anomaly of 0.3 ppb (Figure 7.40). It occurs within a coal bed about 8 m thick. In the first report on this locality, Sweet *et al.* (1990) discussed "range truncations" to within 10 m below the boundary and attributed them to an assumed sensitivity of high-latitude floras to climate change, such as the late Maastrichtian warming proposed by Wolfe and Upchurch (1987b). Despite the suggested comparison with the Wolfe and Upchurch study, no paleobotanical studies were conducted at the Police Island locality. Sweet *et al.* (1990) attributed sporadic occurrences of Maastrichtian species within 2.0–0.2 m below the boundary to unspecified "local environmental factors." Sweet and Braman (2001) provided an extensive review of occurrence data from the Police Island locality (Figure 7.40). They based their analysis on occurrence records of 183 palynofloral species in strata deposited from the mid Maastrichtian to the mid Paleocene. Their summary diagram (Figure 7.40) omits isolated occurrences and certain unspecified long-ranging species of generalized morphology, and they cautioned readers that the diagram cannot be used to determine an extinction percentage. They estimated that about 25% of last appearances were attributable to the K–T boundary event. Because their omission of taxa having only one occurrence within the section is the same method employed by Wilf and Johnson (2004) to obtain a realistic estimate of extinction percentage, it is tempting to use the Sweet and Braman diagram to make an estimate, despite their advice. For the uppermost 10 m below the K–T boundary, that number appears to be well over 50%, but it may be elevated because of the omission of the unspecified long-ranging species, whose inclusion would lower the apparent number. The raw and presumably less detailed occurrence data published by Sweet et al. (1990) indicate an extinction of about 33% at the Police Island locality. An extinction of 25–33% is similar to the magnitudes estimated for the Williston and Powder River basins (see Sections 6.2, 6.3, and 7.4). Irrespective of these facts, Sweet and Braman (2001) concluded that the effect of the terminal Cretaceous impact event on plants was not of sufficient magnitude to override effects of climate changes at the end of the Cretaceous.

7.8   Summary: the record in western North America

Taken together, the palynological and paleobotanical records from 40 well-documented K–T boundary localities in nonmarine rocks in western North America (those having scores of 10 or more in Table 2.1) reveal a clear pattern of regional devastation of the terrestrial vegetation and major extinction of angiosperms in latest Maastrichtian time and gradual recovery of a depauperate flora in the earliest Paleocene (a noteworthy exception to the pattern of gradual recovery is the Castle Rock flora of the Denver Basin). Other K–T boundary localities identified by extinctions of microfossil taxa in association with iridium anomalies are described along the US coast of the Gulf of Mexico and the Atlantic coastal plain (see pertinent papers in Silver and Schultz 1982, Sharpton and Ward 1990, Ryder et al. 1996, and Koeberl and McLeod 2002). We do not discuss these records in this book because the fossils are marine organisms, not terrestrial plants, but those localities complete the record for the continent. As discussed, many of the well-documented K–T boundary localities in North America have yielded geochemical and mineralogical evidence of an extraterrestrial impact in conjunction with the evidence of plant extinctions.

The Chicxulub crater on the Yucatan Peninsula of Mexico is generally accepted as the impact site. We acknowledge late Maastrichtian climate change (see Chapter 5), and cite megafossil paleobotanical data documenting it (Chapters 6 and 7). However, we regard the abrupt extinction of one fourth to one third of palynological taxa as an event of major significance in the history of plants. That event clearly stands out above disappearances attributable to climate change or local variations in sedimentary environment. To evaluate whether or not the K–T boundary event was global in its effects, we turn to records in other regions of the world.

8

Eurasia

8.1 Overview

The quality of published K–T boundary terrestrial data diminishes dramatically outside of North America. Of the 20 non-North American K–T boundary sections, only one scored as much as 10 points on Table 2.1. Our evaluation of these sections begins with the 14 localities known from Eurasia, which we subdivide based on the data available into Europe, Japan, China, and the Russian Far East. Each of these regions is discussed in its own section in this chapter.

Two of the first three places in the world where an iridium anomaly was found at the K–T boundary are in Europe. However, detailed paleobotanical and palynological records of the event are absent at those localities because the boundary occurs in marine rocks. The microfossils that record extinction at the K–T boundary in those places are marine foraminifera, and although they provide an excellent fossil record of the event, they reveal nothing about land plants and the boundary. There are some nonmarine rocks spanning the boundary in Europe, however, and we review the published records from those areas. Abundant nonmarine rock sequences are to be found in Asia, especially in northeast China and the Russian Far East, and extensive literature is available. We have supplemented the Asian literature with our own field work.

A comparison of palynomorph assemblages from Upper Cretaceous and lower Paleogene intervals in western Europe, northwest Africa, and southeast China by Song and Huang (1997) provides an overview of the distribution of plants in those areas at that time, based on palynological records published through 1996 (see also Chapter 5). Song and Huang's data show that during K–T boundary time, western Europe was in the Normapolles Province and at least part of China was in the *Aquilapollenites* Province.

8.2   Europe

As discussed in Chapter 5, at the end of the Cretaceous, western Europe was in the Normapolles palynofloristic province, which extended from eastern North America to western Asia. Floristic changes across the K–T boundary in this broad region were not pronounced. Song and Huang (1997) invoked a change in climate towards cooler conditions from Maastrichtian to early Paleocene time in the Northern Hemisphere to account for changes that did occur. As noted in Chapter 5, however, there is little or no solid evidence for this alleged climate change. Typical Normapolles pollen was common in the Cretaceous in Europe. In the Paleogene, closely related new species appeared that have simpler apertures; these pollen types are collectively known as the "Postnormapolles" group. Thus, in Europe there is a shift in diversity and abundance of Normapolles taxa below the boundary to Postnormapolles taxa above (Song and Huang 1997). This shift cannot be interpreted as an extinction event; it is a gradual change in the composition of successive floras, not an abrupt disappearance of one with replacement by another. On closer inspection, however, it is evident that few critical studies of plants at the K–T boundary have been conducted in Europe, so it is perhaps premature to interpret the record as being in strong contrast with that from North America. Several examples from the western Mediterranean region support this assertion.

Ashraf and Erben (1986) studied palynological successions near the Mediterranean coasts of northeastern Spain and southern France (Figure 8.1). Their database consisted of 224 samples from seven sections, but among these, only one in Spain (Coll de Nargo, locality 86) and two in France (Rousset, locality 89, and Albas, locality 90) span the K–T boundary. They recorded occurrences of 11 species of marine dinoflagellate cysts along with 77 species of terrestrial spores and pollen, and compiled a composite range chart too large to reproduce here. We summarize their data as follows. Almost 80% of the species listed on the chart are pteridophyte spores, and only 12% are angiosperms. A single species of angiosperm (not a Normapolles type) occurs in proximity to the posited position of the boundary, and it is present both below and above it. Independent age control from plant fossils, vertebrates, paleomagnetism, or isotopic ages is lacking at these localities, except for Rousset, where dinosaur eggshells have been reported in the Maastrichtian part of the section. These data do not constitute a useful palynological analysis of the K–T boundary, and Cojan (1989) suggested that in fact the section studied by Ashraf and Erben (1986) did not cross the K–T boundary, but was instead within the Campanian–Maastrichtian transition.

Not far from the study area of Ashraf and Erben (1986), Médus et al. (1988) investigated three nonmarine sections in the Spanish Pyrenees, in the provinces

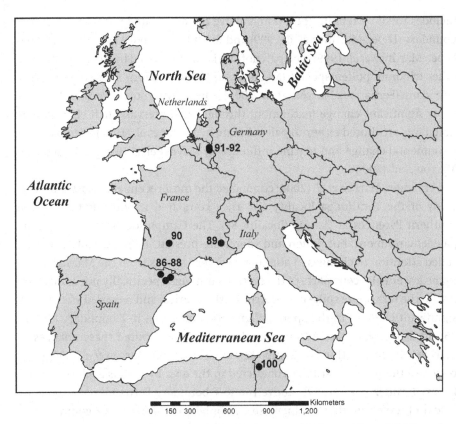

Figure 8.1 Map of western Europe showing approximate positions of K–T boundary localities discussed in the text (numbers are keyed to Table 2.1 and Appendix).

of Barcelona and Lérida (Lleida). Their multidisciplinary study involved palynomorphs, charophytes, ostracodes, and mollusks, and they collected samples for iridium analysis. No iridium anomalies were detected, however. They determined that ostracodes and mollusks had little biostratigraphic value, and few stratigraphic levels proved to be favorable for palynology. Médus *et al.* (1988) concluded that they could not locate the K–T boundary with any precision, but that the K–T "transition" could be recognized from occurrences of charophytes in one of their sections and of palynomorphs in another (Fontllonga). This study is of virtually no use in revealing the nature of the plant record across the K–T boundary in Europe, but subsequent investigations revealed more data.

The section at Fontllonga in the Spanish Pyrenees (locality 87) was the subject of subsequent studies by Médus *et al.* (1992), López-Martínez *et al.* (1998), López-Martínez *et al.* (1999), Mayr *et al.* (1999), and Fernández-Marrón *et al.* (2004). The position of the K–T boundary was determined using magnetostratigraphy (López-Martínez *et al.* 1998). In addition to palynomorphs, the Fontllonga section includes fossil vertebrates below and above the designated position of the K–T boundary (López-Martínez *et al.* 1999) and fossil leaves above (Médus *et al.* 1992, López-Martínez *et al.* 1999). The palynoflora is characterized by Normapolles and other triporate pollen species (Médus *et al.* 1992), and fern spores are notably abundant (Fernández-Marrón *et al.* 2004). The palynomorph record does not show significant change (extinction) through this interval. Such change as is noted was attributed either to paleoclimate by Médus *et al.* (1992) or to paleoenvironmental change and marine influence by Mayr *et al.* (1999) and Fernández-Marrón *et al.* (2004).

Fernández-Marrón *et al.* (2004) conducted the most recent and most thorough study of the Fontllonga locality and they compared it with another in the southern Pyrenees at Campo (locality 88). The Campo locality is in a paralic paleoenvironment, but pollen and spores are present. Fernández-Marrón *et al.* called attention to changes in abundance of fern spores from the Maastrichtian to the lower Paleocene parts of the section (although specifically not comparing it with the fern-spore spike of western North America) and recorded statistically significant changes in the spore-pollen ratio across the K–T boundary (59% to 45% in diversity, 62% to 37% in abundance). They attributed these changes to marine influence within the Campo section. Fernández-Marrón *et al.* (2004) also reviewed the previous studies conducted in the area by Médus *et al.* (1992) and López-Martínez *et al.* (1999). Above the K–T boundary Fernández-Marrón *et al.* (2004) observed neither an increase in abundance of bisaccate gymnosperm pollen as recorded by Médus *et al.* (1992) nor a decrease in abundance of Normapolles pollen as recorded by López-Martínez *et al.* (1999). Thus, available evidence from the western Mediterranean region is not definitive, but it suggests that there was no profound change in palynofloras across the K–T boundary.

In Germany and adjacent parts of central Europe (Figure 8.1), Knobloch *et al.* (1993) reviewed records of occurrence of megaspores, the mesofossils *Costatheca* and *Spermatites*, seeds, and fossil fruits in rocks of Cenomanian to Paleocene age. They found little evidence of abrupt floristic changes from the Maastrichtian to the Paleocene, although many heterosporous plants (known from their megaspores) became extinct by the end of the Cretaceous. Knobloch *et al.* concluded that angiosperms evolved without notable interruption through the K–T transition and there was no significant event at the K–T boundary in central Europe. However, the coarse scale at which they grouped their data (single assemblages from the Maastrichtian and lower Paleocene, respectively) renders this study inconclusive.

Of greater interest are results from a study by Brinkhuis and Schiøler (1996) at Geulhemmerberg, in the southeast part of the Netherlands (Figure 8.1). The

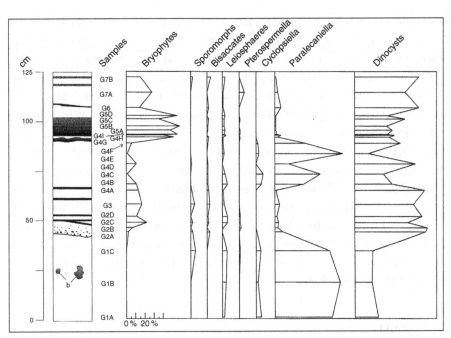

Figure 8.2 Diagram of occurrences of palynomorphs in the K–T boundary interval in the Geulhemmerberg caves, the Netherlands (modified from Brinkhuis and Schiøler 1996). An anomalous abundance of bryophyte spores above the boundary reflects a major change in the ecosystem and may be analogous to a fern-spore spike. The K–T boundary is placed just below sample G2A, which is at the level where bryophyte spores first appear. Reprinted with kind permission of Springer Science and Business Media.

depositional environment at Geulhemmerberg (locality 91) was marginal marine, inner neritic, with a substantial influx of terrestrial palynomorphs. Brinkhuis and Schiøler determined the position of the K–T boundary from occurrences of marine dinoflagellate cysts. At the boundary they noted an anomalous increase in the relative abundance of bryophyte spores (Figure 8.2) that may indicate a major change in the terrestrial ecosystem. They surmised that the abundance of bryophyte spores might be analogous to the fern-spore spike. Their data are based on 24 samples from an interval of 125 cm in which palynomorphs are well preserved. Unfortunately, other than bryophyte spores, terrestrial palynomorphs are exceedingly scarce in these samples.

The final study of the K–T boundary in Europe that we judge to be of relevance to the record of plants is that of Herngreen *et al.* (1998), also in the Netherlands. These authors investigated the biostratigraphy of foraminifera, ostracodes, calcareous nannofossils, and palynomorphs (including dinoflagellates, spores, and pollen) in the Maastrichtian–Danian (Maastrichtian–lower

Paleocene) interval in Curfs Quarry (locality 92), in the Maastrichtian type area. Recovery of spores and pollen from marine limestone samples was low. Assemblages of pollen clearly belong to the Normapolles Province. Herngreen et al. (1998) concluded that there is no indication of drastic change in the character or composition of terrestrial vegetation at the K–T boundary in the Curfs Quarry section. They attributed the changes they did record to paleoclimate. Their diagram of relative abundances of palynomorphs in their section (Figure 8.3) shows a sudden increase in the abundance of bryophyte spores in the Paleocene similar to that reported by Brinkhuis and Schiøler (1996).

The significance of the bryophytes in the Dutch sections is unclear, and it could be attributed to changes in paleoenvironment if not paleoclimate. In summary, the record from Europe is sparse, ambiguous, and non-definitive. There seems to be little or no evidence of an extinction event among angiosperms, but too few detailed studies have been conducted in nonmarine strata to build a reliable database.

8.3   Japan

The islands of Japan lie within the *Aquilapollenites* Province (see Chapter 5), and hence palynological assemblages from the Maastrichtian and Paleocene rocks in Japan have much in common with those of western North America. Whereas the history of plants at the K–T boundary in the Normapolles Province of Europe is ambiguous, we would expect the record from Japan to be easier to interpret, because of the floristic similarities with North America. However, detailed studies relating to the boundary have not been conducted. The most intriguing records are from Hokkaido, the northernmost of the Japanese islands (Figure 8.4). Palynologist Kiyoshi Takahashi conducted the studies that provide most of the data from this area, beginning in the early 1960s. Unfortunately, most of these reports are descriptions of palynofloras of Late Cretaceous age from various stratigraphic units in Japan and do not pertain to the K–T boundary.

Takahashi and Shimono (1982) described a palynoflora from west-central Japan that includes 114 species, 75 of which are angiosperm pollen. Among the angiosperms are 43 species Takahashi and Shimono assigned to the Triprojectacites group. As explained in Section 5.1, this group of morphologically distinctive pollen includes the genus *Aquilapollenites* and, according to Takahashi and Shimono, 10 other genera, many of which we treat as synonyms of *Aquilapollenites* (Figure 8.5). Our taxonomic viewpoint aside, Takahashi and Shimono described 34 species that are either in the genus *Aquilapollenites* or in closely related genera. They used this extraordinary assemblage to determine

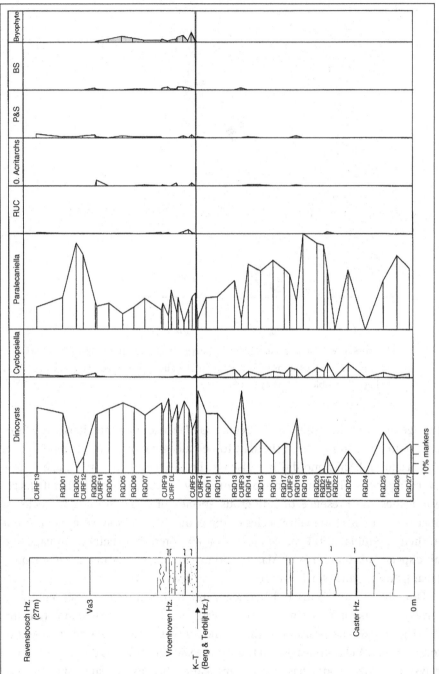

Figure 8.3 Diagram of occurrences of palynomorphs in the K–T boundary interval in Curfs Quarry, the Netherlands (from Herngreen et al. 1998). The anomalous abundance of bryophyte spores (column at far right, shaded for emphasis) may be analogous to a fern-spore spike. Other columns show relative abundances of dinoflagellate cysts, acritarchs, and other pollen and spores. Reprinted by permission.

Figure 8.4 Map of East Asia including Japan, China, and the Russian Far East show-ing approximate positions of K–T boundary localities discussed in the text (numbers are keyed to Table 2.1 and Appendix).

the age of the rocks from which it came as Maastrichtian; previously it was of uncertain age and had been thought to be Paleogene or even Neogene in age. Takahashi and Shimono did not discuss the K–T boundary or the age of any overlying rocks. The importance of their study for our purposes is that it clearly establishes the presence of numerous species of *Aquilapollenites* in rocks of Maastrichtian age in Japan. Species of the genus *Wodehouseia* were also present in their assemblage. Takahashi and Shimono noted that their assemblage had nine species in common with Maastrichtian assemblages in Hokkaido (northern Japan), Siberia, and North America.

Directly relevant to the K–T boundary in Japan, Saito *et al.* (1986) reported a fern-spore spike in a marine section, Kawaruppu in eastern Hokkaido (locality 93); Figure 8.6. The position of the boundary was determined on the basis of foraminifera. A claystone layer in the section was identified as the K–T boundary claystone. This report has great significance because it suggests that the

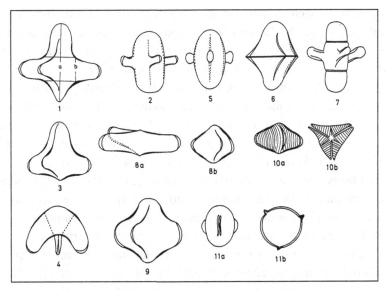

Figure 8.5 Diagrams illustrating triprojectate pollen including *Aquilapollenites* and others considered synonymous with that genus (1–7) from the *Aquilapollenites* palynofloral province (from Takahashi and Shimono 1982). Other genera illustrated are *Fibulapollis* (8a, 8b), *Pentapollenites* (9), *Cranwellia* (10a, 10b), and *Orbiculapollis* (11a, 11b). Reprinted by permission.

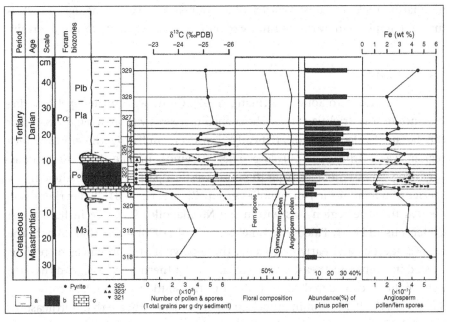

Figure 8.6 K–T boundary interval at a marine locality in Hokkaido, Japan (modified from Saito *et al.* 1986). A possible fern-spore spike is present, as indicated by the dashed line in the column at far right. Reprinted by permission.

fern-spore spike, well known in western North America, may be present in Japan, as well. With regard to the putative K–T boundary claystone, no evidence verifying its identification, such as the presence of an iridium anomaly or shocked quartz, was presented. Other than the fern-spore abundance, there is little palynological data from the section. Only the relative abundance of gymnosperm pollen is discussed (it increases above the boundary); there are no data on the stratigraphic distribution of pollen taxa, such as *Aquilapollenites* or associated genera. We do not question that the position of the K–T boundary was correctly identified on the basis of foraminifera; the details of the foraminiferal record were published the same year by Kaiho and Saito (1986). We regard this locality as one that might benefit from further palynological study, but in fact a later study in the same area by Takahashi and Yamanoi (1992) produced ambiguous results.

Takahashi and Yamanoi (1992) analyzed the palynofloristic changes through the Cretaceous–Paleogene transition in the Katsuhira Formation of eastern Hokkaido. According to Saito *et al.* (1986), this interval spans the K–T boundary. No significant palynofloristic change was observed within the interval, however. Takahashi and Yamanoi acknowledged and reviewed the North American record, but they did not recognize a parallel change within the Katsuhira Formation.

Thus, the plant microfossil record of the K–T boundary in Japan, although intriguing, remains unclear. A major obstacle to clarification may be the lack of nonmarine rocks of latest Maastrichtian and earliest Paleocene age, which is further complicated by discontinuity of outcrop in the heavily forested terrain.

8.4   China

According to Song and Huang (1997), China was in the *Aquilapollenites* Province in Maastrichtian time. China encompasses a vast area (Figure 8.4) that can be subdivided palynofloristically, however, and only the northeastern region was within the *Aquilapollenites* Province in the Late Cretaceous. Palynological assemblages from the northwestern part of China include many species of pollen of the Normapolles group. Zhao *et al.* (1981) had earlier documented that the region was within the Normapolles Province in both latest Cretaceous and early Paleogene time. Wang *et al.* (1990) stated that, in both the Late Cretaceous and early Paleogene, three palynofloral provinces or regions existed in China: northeastern, western, and southern (they assigned the area that might be considered central China to the southern region). According to Song *et al.* (1983), the dominant pollen types present indicate that northeastern China had a semi-humid, subtropical to temperate climate in the Late Cretaceous, and a humid warm-temperate climate in the Paleocene; the rest of

China (western, central, and southern) had an arid, subtropical climate in both Late Cretaceous and Paleocene time.

Several summaries of Upper Cretaceous nonmarine stratigraphy within various basins and provinces of China have been published that included some Paleocene units (Chen 1983, Hao and Guan 1984, Chen 1996, Chen et al. 2006). Most of the biostratigraphic determinations are based not on plant fossils but on conchostrachans, ostracodes, and mollusks. Disparities in indicated ages and correlations among these summaries suggest considerable uncertainty in the exact ages and stratigraphic relations of the units. Mateer and Chen (1992) reviewed potential K–T boundary intervals in nonmarine rocks of China. They reported that the paleontological records of most of the ten basins they investigated primarily involve charophytes and ostracodes rather than plant fossils. They concluded that, based on palynomorphs, potential K–T boundary localities may exist in the Minghe Basin in Gansu and Qinghai provinces (western region), the Jiangsu Basin in Jiangsu Province (southern region), and the Songliao Basin in Heilongjiang and Liaoning provinces (northeastern region). Here we discuss the data available from each of these regions. The southern and northeastern regions have yielded the most data.

Wang et al. (1990) briefly summarized data on palynological assemblages of Cenomanian to Miocene age from Xinjiang and Qinghai provinces, which are, respectively, in the western and southern palynofloral regions they defined. They were unable to locate the K–T boundary in these regions; they found mixed assemblages in both of their study areas.

Wang and Zhao (1980) described four palynological assemblages from the Jianghan Basin in Hubei Province, among which the third encompasses the upper part of the Cretaceous (Senonian) through the lower part of the Paleocene. Hubei Province evidently lies within the southern palynofloral region of Wang et al. (1990). The palynofloras of this region are characterized by an abundance of ephedralean gymnosperm pollen species, and Wang and Zhao (1980) concluded that the flora indicates an arid climate. Their paper included a range chart of selected species grouped into the four assemblages; it shows no break within the stratigraphic distribution of the third assemblage, and hence the K–T boundary is not distinguishable.

A Maastrichtian and Paleocene section in the Nanxiong Basin, Guangdong Province (locality 94), has been the subject of extensive studies by several authors, most notably Zhao et al. (1991), Erben et al. (1995), Stets et al. (1996), Buck et al. (2004), and Taylor et al. (2006). Guangdong Province is in the southern palynofloral region of China. The stratigraphic units of interest are the Pingling Formation of the Nanxiong Group and the Shanghu Formation of the Luofozhai Group. These are primarily red beds, which are notoriously poor for palynomorph recovery. Stets *et al.* (1996) discussed collecting and preparing 425 samples, but without being specific they noted that the recovery of palynomorphs was less than they had anticipated, which is probably an understatement. Nonetheless, they were able to identify 41 palynomorph species, most of which are pteridophyte spores. They identified seven species of gymnosperm pollen, none of which included the ephedralean types that according to Wang and Zhao (1980) are characteristic of the southern palynofloral region of China. On the basis of their palynostratigraphy, Stets *et al.* claimed to have bracketed the boundary within an interval 21 m thick, 80 to 101 m below the top of the Pingling Formation. A previous investigation (Zhao *et al.* 1991) had placed the K–T boundary at the contact of the Pingling and the overlying Shanghu formations, primarily on a lithostratigraphic basis. Stets *et al.* (1996) reported extinction of the majority of taxa they regarded as indicative of Late Cretaceous age. Neither a K–T boundary claystone layer nor an iridium anomaly was found, but Stets *et al.* noted changes in the ratios of the stable isotopes ^{18}O and ^{13}C. They interpreted that their stratigraphic section is within an interval of reversed geomagnetic polarity, which they assumed to be subchron C29r. Stets *et al.* reported fragments of dinosaur eggshells above their bracketed position of the K–T boundary, in the uppermost part of the Pingling Formation. They concluded that these eggshell fragments are evidence that dinosaurs had survived into the Paleocene. Although this book is not about dinosaurs and the K–T boundary, this report of dinosaur fossils (eggshell fragments) above a putative K–T boundary established by palynology requires further comment.

Dinosaur eggshell fragments and nests containing complete eggs were first reported in the Nanyung [Nanxiong] Group by Young [Yang] and Chow [Zhou] (1963). Zhao *et al.* (1991) also discussed the presence of eggshell fragments and used their highest stratigraphic occurrence to support placing the K–T boundary well above the position designated by Stets *et al.* (1996). The associated paleomagnetic data led Zhao *et al.* (1991) to conclude that dinosaur extinction in southern China took place 200 000–300 000 years before the terminal Cretaceous event. This deduction did not go unchallenged, however. Russell *et al.* (1993) published a reinterpretation of the paleomagnetic data, concluding that a hiatus representing more than six million years exists at the contact between the Pingling Formation and the Shanghu Formation, where Zhao *et al.* (1991) placed the K–T boundary. Russell *et al.* also noted that no iridium anomaly that could verify the position of the boundary had been found in the section studied by Zhao *et al.*

Zhao *et al.* (2002) later reported minor iridium anomalies at three different levels within the Pingling Formation, based on analyses of the dinosaur eggshell fragments (none of these "anomalies" is associated with a boundary claystone layer). Stets *et al.* (1996), who placed the K–T boundary 80 to 101 m below the top of the Pingling Formation, had concluded that dinosaur eggs and eggshell fragments above that level were Paleocene in age.

Interpretations of the stratigraphic occurrence and biostratigraphic significance of dinosaur eggshells and the placement of the K–T boundary in the Nanxiong Basin were challenged by Buck *et al.* (2004). Buck *et al.* restudied the "Nanxiong" [Pingling] Formation and the overlying Shanghu Formation. Utilizing the palynologic data of Stets *et al.* (1996), Buck *et al.* (2004) concurred with Stets *et al.* in placing the K–T boundary in the upper part of the Pingling Formation, well below its contact with the overlying Shanghu Formation. Based on their interpretations of lithofacies, Buck *et al.* (2004) concluded that the uppermost 101 m of the Pingling Formation and the lowermost part of the Shanghu Formation were deposited as debris flows that reworked Cretaceous fossils, including the dinosaur eggshell fragments, above the level of the K–T boundary. Buck *et al.* concluded that the assertions of Zhao *et al.* (1991) and Stets *et al.* (1996), that dinosaur fossils occur above the K–T boundary in the Nanxiong Basin, is based on a misunderstanding of the sedimentological origin of the deposits in the upper part of the Pingling Formation and the lower part of the Shanghu Formation.

The most recent and most comprehensive study of the controversial Pingling–Shanghu red-bed section in the Nanxiong Basin is that of Taylor *et al.* (2006). These authors reviewed all of the pertinent literature and revisited the site. Their results are presented in Figure 8.7. On the basis of the data from vertebrate paleontology including dinosaur eggs and Paleocene mammals, they reasserted that the placement of the K–T boundary at the contact between the Pingling and Shanghu formations established by Zhao *et al.* (1991) is correct. As for the claim by Buck *et al.* (2004) that the uppermost part of the Nanxiong [Pingling] Formation is composed of debris-flow sediments containing reworked fossils, Taylor *et al.* (2006) noted the presence of *in situ* dinosaur eggs in nests, which could not have been reworked, right below the formational contact. Taylor *et al.* did not review the palynological data of Erben *et al.* (1995), Stets *et al.* (1996), and Zhao *et al.* (2002), which Buck *et al.* (2004) had accepted. Our review of those data leaves us skeptical that any of the species of fossil pollen said to be indicative of Tertiary age are exclusively Paleogene. We conclude that the position of the K–T boundary in the Nanxiong Basin remains unresolved, but if the well-reasoned interpretations of Taylor *et al.* (2006) are correct, the evidence for Paleocene dinosaurs is unconvincing.

Returning to the data on plants and the K–T boundary in China, we move to the northeastern part of that country. In the earliest published report on the K–T boundary in China, Hao *et al.* (1979) studied the Mingshui Formation in the Songliao Basin, a large geologic feature in the northeastern part of the country

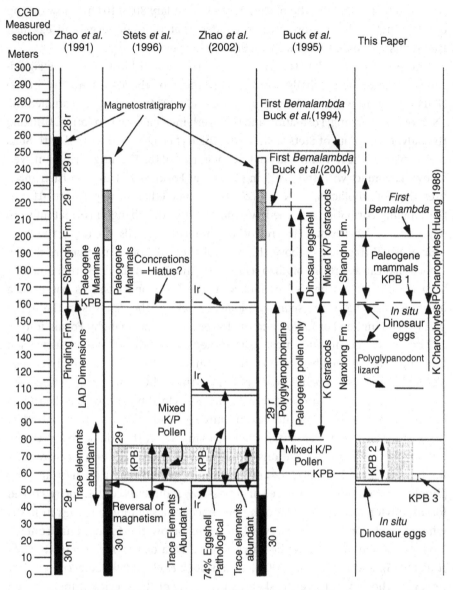

Figure 8.7 The K–T boundary interval in the Nanxiong Basin, Guangdong, China (modified from Taylor *et al.* 2006). Various possible stratigraphic positions of the K–T boundary (KPB) selected in previous studies are indicated in the first four columns and summarized in the fifth; Taylor *et al.* favored KPB 1 in the fifth column. Reprinted by permission.

that occupies parts of Heilongjiang, Jilin, Nei Mongol, and Liaoning provinces. They reported that an assemblage with *Aquilapollenites* and other triprojectate pollen, "*Proteacidites*" (= *Tschudypollis*), and *Wodehouseia* is present in the lower part of the formation. The assemblage in the upper part of the Mingshui includes abundant *Deltoidospora* and *Ulmipollenites*. They noted that the lower assemblage in the Mingshui Formation is comparable to that in the Lance Formation of Wyoming and the upper assemblage is comparable to that in the Fort Union Formation of the United States. Based on these observations, Hao *et al.* (1979) placed the K–T boundary between the lower and upper parts of the Mingshui Formation. Based on the data they presented, we concur with that interpretation. Hao *et al.* also called attention to similarities between the assemblages in the Mingshui Formation and those in the Tsagayan Formation of the Russian Far East (that formation will be discussed at length in Section 8.5).

Song *et al.* (1980) briefly summarized palynological assemblages of Aptian through Miocene age in Jiangsu Province, which lies within the *Aquilapollenites* palynofloral province in northeastern China. The K–T boundary apparently is within the Taizhou Formation in Jiangsu Province. There the disappearance of species of *Aquilapollenites*, "*Proteacidites*" (*Tschudypollis*), and *Wodehouseia* was reported, along with an increase in conifer pollen said to reflect climatic cooling. The authors noted that the palynofloral change corresponds to the transition from the lower to the upper member of the Taizhou Formation, but see the discussion of the later work of Song *et al.* (1995), below.

Song *et al.* (1995) revisited the Maastrichtian and Paleocene of the Taizhou Formation in northern Jiangsu. They defined two assemblages within the formation; the lower one is divided into three sub-assemblages, and the upper one is divided into two sub-assemblages (Figure 8.8). They concluded that Jiangsu Province lies at the border of the Normapolles and *Aquilapollenites* palynofloral provinces because the Cretaceous sub-assemblages contain species characteristic of both. It must be noted, however, that these authors followed Song *et al.* (1983) in recognizing only two palynofloral provinces in China – the northeastern one with *Aquilapollenites* and a central one that includes both the southern and western regions, which they assigned to the Normapolles Province. The assemblage Song *et al.* (1995) described and illustrated from the lower part of the Taizhou Formation includes many of the ephedralean gymnosperm pollen species reported by Wang and Zhao (1980) and Wang *et al.* (1990), which characterize a distinctive palynofloral province in southern China. This disagreement about the presence or absence of the Normapolles Province in China notwithstanding, a change in palynofloras within the upper part of the Taizhou Formation was interpreted by Song *et al.* (1995) as marking the transition from the Maastrichtian to the Danian (lower Paleocene).

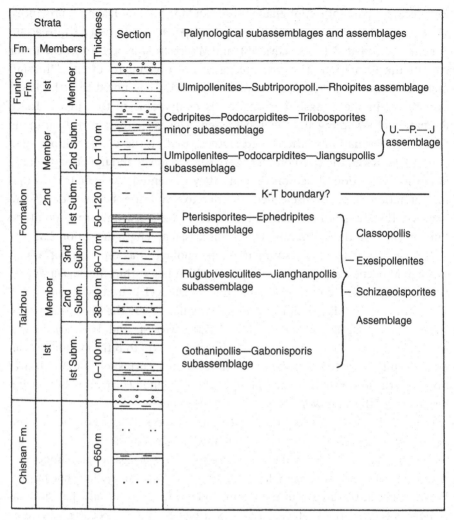

Figure 8.8 The K–T boundary interval in Jiangsu Province, China, with a possible approximate position of the boundary indicated by the horizontal line and question mark (modified from Song *et al.* 1995). See text for discussion. Reprinted by permission of Elsevier.

The Danian or Paleocene age of the uppermost sub-assemblage of the Taizhou Formation proposed by Song *et al.* (1995) is based primarily on the increased relative abundance of conifer pollen, ostensibly by comparison with records from North America. If the North American record were to be invoked as a comparison, however, we would place the K–T boundary in the Taizhou Formation at a position stratigraphically lower than that favored by Song *et al.* (1995), based on the disappearance of *Aquilapollenites*, *Proteacidites*, and

Wodehouseia mentioned by Song *et al.* (1980). We tentatively suggest the approximate position of the K–T boundary shown in Figure 8.8. An increase in the abundance of conifer pollen may well be a paleoclimatic signal, as interpreted by Song *et al.* (1995), but we doubt that it is a useful indicator of the position of the K–T boundary. These results are further evidence that sections without iridium are difficult to interpret.

Heilongjiang Province is in the far northeastern part of China (Figure 8.4) and lies within the *Aquilapollenites* palynofloral province. In a primarily taxonomic study, Liu (1983) described the Maastrichtian and Paleocene palynofloras of the Furao and "Wuyin" [Wuyun] formations in Heilongjiang, listing 215 species in 119 genera and illustrating 180 of those species. Liu implied that the K–T boundary is within the Furao Formation in the Heilongjiang area and indeed several species that he illustrated from the lower part of the formation are similar to, or the same, as those in the Maastrichtian of western North America. The overlying Wuyun Formation carries some species also known in western North America and evidently is Paleocene in age, although Liu noted that some species of *Aquilapollenites* are still present, a pattern unlike that in North America.

Sun *et al.* (2002) investigated possible K–T boundary sections along the Heilongjiang River in Heilongjiang Province including locality 95 of Table 2.1 and the Appendix. One of us (KRJ) was a member of that field symposium. The party discovered leaf megafossils at several stratigraphic levels in the Yong'ancun, Taipinglinchang, and Wuyun formations and visited hadrosaur bone beds in the Yuliangzi Formation (Figure 8.9). Johnson discovered a hadrosaurian dinosaur footprint in the Upper Cretaceous Yong'ancun Formation. The Taipinglinchang megaflora contained abundant leaves of *Cobbania corrugata* and *Quereuxia angulata*, aquatic taxa that co-occur in Campanian and Maastrichtian rocks in western North America. Leaf fossils were collected from the Wuyun Formation exposed in a coal mine near the town of Wuyun. Sun *et al.* (2002) named a new stratigraphic unit, the "Baishantou Member" of the Wuyun Formation (Figure 8.9). The Baishantou Member was separated from the Furao Formation at a tuff bed marker unit at its base and assigned to the Wuyun Formation. The megafossil assemblage from the Baishantou Member is dominated by *Tiliaephyllum tsagajanicum* and was believed to be early Paleocene in age by Russian paleobotanists who were also members of the field symposium. The overlying coal-bearing member of the Wuyun Formation yielded abundant fossil leaves including *"Ampelopsis" acerifolia*, *Corylites*, and *Cercidiphyllum* that are strikingly similar to middle Paleocene megafloras from Wyoming (Gemmill and Johnson 1997). Thus, based on a North American

Series and Stages			RUSSIA			CHINA	
			Zeya-Bureya Basin			Heilongiang Province	
Paleocene	Danian	U	Tsagayan Group	Darmakan Formation	coal-bearing subformation	Wuyun Formation	coal-bearing member
		L			sandstone subformation		Baishantou Member
Upper Cretaceous	Maastrichtian	U		Bureya Formation			Furao Formation
		M		Udurchukan Formation (main dinosaur horizon)			Yuliangzi Formation (main dinosaur horizon)
		L					
	Campanian		Kundur Formation		Upper Zavitan Formation	Taipinglinchang Formation	
	Santonian						
	Coniacian		Boguchan Formation			Yong'ancun Formation	
	Turonian						
	Cenomanian		Lower Zavitan Formation			?	

Figure 8.9 Stratigraphic nomenclature for the Upper Cretaceous and Paleocene of Heilongjiang Province, northeastern China (column at right), showing correlations with the stratigraphically equivalent formations and members in the Russian Far East. The lower and upper contacts of the members of the revised Wuyun Formation are not defined lithostratigraphically (modified from Sun *et al*. 2002). Reprinted by permission.

biostratigraphy, the K–T boundary would appear to lie below the Wuyun Formation and above the Yuliangzi Formation, i.e. within the Furao Formation (Figure 8.9).

As noted, Liu (1983) had implied that the K–T boundary in Heilongjiang Province lies within the Furao Formation. Sun *et al*. (2002, 2004) proposed that the type area of the Furao Formation should be restudied because it may actually be composed of a lower part of latest Cretaceous age and an upper part of early Paleocene age. Sun *et al*. reasoned that the K–T boundary in the Heilongjiang area lies within an interval below the base of their Baishantou Member, in the uppermost part of the Furao Formation. In 2004, Dr. Sun Ge of the Research Center of Paleontology and Stratigraphy, Jilin University, Changchun, collected six samples for palynological analysis from the Baishantou Member, which he sent to one of us (DJN) for analysis. Results confirmed an early Paleocene age for the leaf bed that contains the *Tiliaephyllum tsagajanicum* assemblage, although microfossil recovery from that bed was poor. Two samples from lower in the section yielded good palynomorph assemblages of early Paleocene age. Unfortunately, three samples from the Furao Formation below the tuff bed were barren, and the possible Late Cretaceous age of that interval could not be verified. Thus, the exact position of the K–T boundary in Heilongjiang Province remained to be determined.

As of 2007, a search for the K–T boundary in northeastern China was under way using samples from two drill cores in the Furao Formation, the Xiaoheyan 2005 and 2006 cores from near Wuyun, Heilongjiang Province. Drilling of the cores was arranged by Sun Ge. Samples collected for palynology are under investigation by an international team including Rahman Ashraf of the Institut für Geowissenschaften, Tübingen, Germany; Ian Harding, University of Southampton, England; Valentina Markevich of the Russian Academy of Sciences, Vladivostok; and one of us (DJN). Of particular interest are species of *Aquilapollenites* present in the samples, some of which are also found in western North America, and others that evidently are indigenous to northeastern China and adjacent areas in the Russian Far East. However, the stratigraphic ranges of species of *Aquilapollenites* and some others that are characteristic of the upper-most Maastrichtian of western North America are not definitively known in northeastern China and the Russian Far East. Recovery from the Xiaoheyan 2005 core was poor and unreliable in some intervals, which necessitated drilling of the Xiaoheyan 2006 core, from which recovery was nearly complete. Based on presence of pollen species in the lower parts of the Xiaoheyan 2006 core that are known to be Maastrichtian in North America, that interval appears to DJN to be Cretaceous in age. Based on stratigraphic ranges of those and some other species in the Russian Far East, Valentina Markevich also concluded (personal communication, 2007) that the lower part of the cored interval is Maastrichtian in age. A sample of uncertain stratigraphic position within the upper part of the Xiaoheyan 2005 core appeared to be of Danian age according to Markevich, but no samples from the upper part of the 2006 core yielded a similar assemblage. The investigations are continuing, but the results from the Xiaoheyan 2006 core, in which the goal was to locate the K–T boundary, verified the presence of the "K" in the lower part of the interval but not the "T" in the upper part.

The sections in northeastern China that include uppermost Cretaceous and lowermost Paleocene nonmarine rocks appear to hold much potential for future discoveries of direct relevance to plants and the K–T boundary. Although initial results from the Wuyun area in Heilongjiang Province were inconclusive, coring seems to be the best way to obtain samples for palynological analysis, because outcrops in the area are discontinuous at road cuts, in coal mines, or along the

Heilongjiang River. At the current state of knowledge, however, sites that rely solely on palynological data and that lack geochronologic, magnetostratigraphic, or geochemical data must be regarded at best as potential K–T boundary sections. At present, the resolution of the K–T boundary in this region is at the stage level (see Section 2.1).

8.5   The Russian Far East

We now cross the Heilongjiang River of China to where that same river is called the Amur, the area known as the Russian Far East. The stratigraphy of the Maastrichtian and Paleocene in the Zeya–Bureya Basin of the Russian Far East (Figure 8.10) is correlative with that of Heilongjiang Province. The distance between these areas is only the width of the Heilongjiang/Amur River but politics has, until recently, prevented the direct comparison of the two adjacent

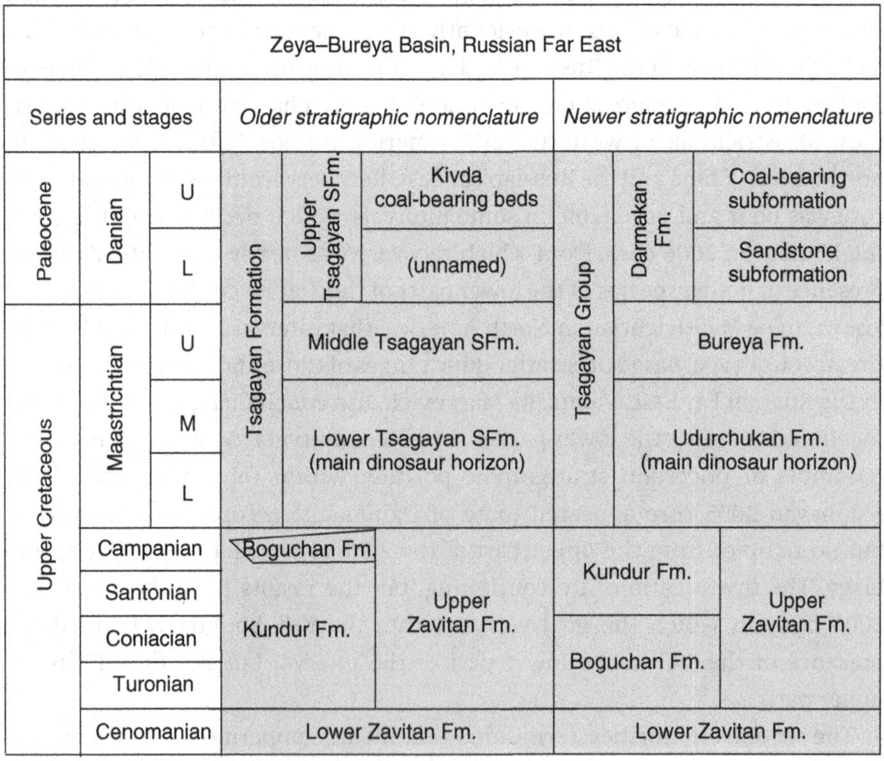

Zeya–Bureya Basin, Russian Far East				
Series and stages			*Older stratigraphic nomenclature*	*Newer stratigraphic nomenclature*
Paleocene – Danian U	Tsagayan Formation – Upper Tsagayan SFm.	Kivda coal-bearing beds	Tsagayan Group – Darmakan Fm.	Coal-bearing subformation
Paleocene – Danian L		(unnamed)		Sandstone subformation
Upper Cretaceous – Maastrichtian U		Middle Tsagayan SFm.		Bureya Fm.
Upper Cretaceous – Maastrichtian M / L		Lower Tsagayan SFm. (main dinosaur horizon)		Udurchukan Fm. (main dinosaur horizon)
Campanian	Boguchan Fm. / Kundur Fm.	Upper Zavitan Fm.	Kundur Fm.	Upper Zavitan Fm.
Santonian				
Coniacian			Boguchan Fm.	
Turonian				
Cenomanian	Lower Zavitan Fm.		Lower Zavitan Fm.	

Figure 8.10 Stratigraphic nomenclature for the Upper Cretaceous and Paleocene of the Amur region, Russian Far East, showing both older and newer names of units; the abbreviation SFm. stands for subformation, a stratigraphic rank used in Russian nomenclature (modified from Sun *et al.* 2002 with data from Geological Institute, Russian Academy of Sciences 2003). Reprinted by permission.

areas. Beginning in 2000, Russian and Chinese workers have begun to collaborate and visit each other's sections (Sun 2002, 2004).

The earliest paper in the Russian palynological literature that directly relates to the K–T boundary is by Mtchedlishvili (1964), in which the author discussed the stratigraphic and geographic distribution of Triprojectacites pollen (*Aquilapollenites* and closely related genera). As discussed, this pollen group constitutes a major component of contemporaneous palynofloras of western North America and eastern Asia.

Mtchedlishvili (1964) stated that the most widely distributed species of *Aquilapollenites* in Russia is *A. quadrilobus* Rouse (which coincidentally is the type-species of the genus). This species is said to occur in various parts of Russia in rocks as old as Cenomanian and as young as early Paleocene. In North America, the occurrence of this species is restricted to the interval from the lower Campanian to the uppermost Maastrichtian (Tschudy and Leopold 1970, Nichols 1994) and it disappears at the K–T boundary. These dissimilar ranges are difficult to reconcile, given the wide distribution of the species on both continents. In Russia, Triprojectacites pollen has been recorded from the Santonian onward, which is in general agreement with the records from North America. The pollen group reached its peak of development in the Maastrichtian, and species of *Aquilapollenites* and related genera are found in rocks of that age in western Siberia, Yakutia, and the Russian Far East, where they are the characteristic forms at all localities. However, Mtchedlishvili also stated that *Aquilapollenites* is represented by even greater numbers of specimens, although fewer species, in rocks of Maastrichtian to Danian (Paleocene) age in western Siberia, and that triprojectate pollen is found as rare, isolated specimens in rocks as young as middle Eocene or early Oligocene. As the Paleogene records are inconsistent with those from North America, such records raise a fundamental question about stratigraphic occurrence data from Russia: how is it independently known that the age of a given deposit is any specific age? For the Cretaceous in most areas of western North America, the answer is that the ages of stratigraphic units are tied to ammonite biostratigraphy and/or radiometric dates (e.g., Nichols and Sweet 1993, Obradovich 1993). Such data are scarce or lacking in Russia, however, and thus we have become wary of the assignment of Russian terrestrial formations to standard Cretaceous or Paleogene stages.

Mtchedlishvili (1964) noted that *Aquilapollenites subtilis* Mtchedlishvili, which is common in the Maastrichtian in Russia, also occurs in the lower Paleocene. Mtchedlishvili observed that this species closely resembles, and may be the same as, the North American species *A. spinulosus* Funkhouser (Figure 8.11). In North America, *A. spinulosus* is unknown in the Maastrichtian and first occurs in

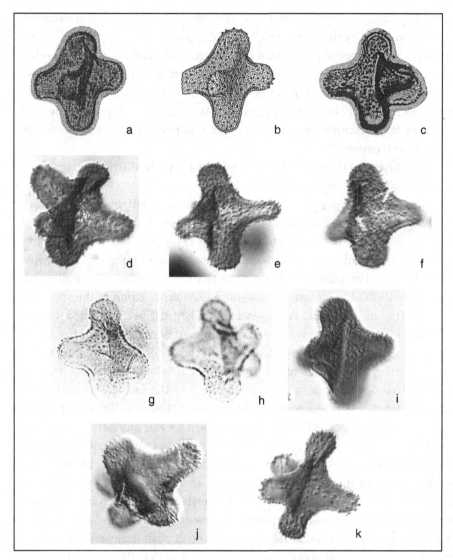

Figure 8.11 *Aquilapollenites subtilis* Mtchedlishvili 1961 (Russia) and *Aquilapollenites spinulosus* Funkhouser 1961 (North America). a – photomicrograph of holotype specimen of *A. subtilis* (from Samoilovich *et al.* 1961, Fig. 3a), b – drawing of holotype specimen of *A. subtilis* (from Samoilovich *et al.* 1961, Fig. 3b), c – photomicrograph of paratype specimen of *A. subtilis* (from Samoilovich *et al.* 1961, Fig. 4), d–f – photomicrographs of specimens of *A. subtilis* (Tsagayan Formation, Russia), g–h – photomicrographs of holotype specimen of *A. spinulosus* (from Funkhouser 1961, Figs. 4a–4b), i – photomicrograph of *A. spinulosus* (Fort Union Formation, North Dakota), j–k – photomicrographs of *A. spinulosus* (Fort Union Formation, Wyoming). All specimens about 30 micrometers in greatest dimension.

the middle Paleocene, after all other species of the genus are extinct. These two species (which may be one in the same) clearly have disparate ranges in eastern Asia and western North America, but migration from Asia to North America during middle Paleocene time could account for the sudden appearance of *A. spinulosus*, which has no direct ancestors in North America.

All species of *Aquilapollenites* disappeared from Russia and North America at some time in the Paleocene. Commenting on the cause of the ultimate extinction, Mtchedlishvili (1964) asserted that there were no data that would identify the factors responsible, but that climate change – in particular a drop in temperature during Paleocene time – may have been among the causes. The possible influence of climate change in the extinctions at the end of the Cretaceous was later reiterated by other Russian palynologists and paleobotanists.

A major early publication with specific reference to the palynology and paleobotany of the Russian Far East is that of Bratseva (1969). Her study area was along the Amur River near the city of Blagoveshchensk and the towns of Raichikhinsk and Arkhara, all of which became familiar to one of us (DJN) during field work conducted more than 30 years later, in 2003. Bratseva (1969) discussed how the age of the rich Tsagayanian megafossil flora in the Tsagayan Formation of the Zeya–Bureya Basin (Figure 8.10) was uncertain. The Tsagayanian flora was thought to be Danian (Paleocene) until palynological analyses conducted in the 1950s suggested that it was older and should be assigned either to the ambiguous "Maastrichtian–Danian" or that it might even be as old as Cenomanian. Bratseva set out to resolve the age dispute.

After analyzing 285 samples from 15 outcrops and 18 boreholes from the Aptian–Albian upward, Bratseva (1969) concluded that the Tsagayanian mega-flora is Maastrichtian in age, and that deposits of Danian age begin with the Kivda coal-bearing beds in the Zeya-Bureya Basin (Figure 8.10). She noted that the palynofloras of the Tsagayan and Kivda beds differ strongly from one another in floristic composition. Triprojectate pollen predominates in the beds bearing the Tsagayanian megaflora, and most of those species are absent from the Kivda beds. Bratseva specifically compared the Tsagayanian palyno-floral assemblage with those from the Lance and Hell Creek formations in the western United States, stating that the species in common confirmed the Maastrichtian age of the Tsagayanian megafloral assemblage. In contrast, the Kivdian palynoflora is composed largely of pollen referable to the modern families Myricaceae, Juglandaceae, Fagaceae, Hamamelidaceae, and other tripo-rate pollen genera, and is similar to assemblages from the Paleocene deposits of the Tullock, Ludlow, and Cannonball members of the Fort Union Formation in

Stage	Palynozone	Bureya Depression
Danian	XIV *Triatriopollenites plicoides–Comptonia sibirica*	Lower Kivdinskaya
Danian	XIII *Wodehouseia fimbriata–Ulmipollenites krempii*	Upper Tsagayanskaya
Maastrichtian	XII *Orbiculapollis lucidus–Wodehouseia avita*	Middle Tsagayanskaya
Maastrichtian	XI *Wodehouseia spinata–Aquilapollenites subtilis*	Lower Tsagayanskaya

Figure 8.12 Palynostratigraphic zonation of the Maastrichtian and Danian of the Russian Far East (modified from Markevich 1994). Reprinted by permission of Elsevier.

North America. Bratseva reported an abrupt and fundamental change in the composition of the palynofloras between the Tsagayan and Kivda assemblages, in which species of *Aquilapollenites* and other triprojectate pollen almost entirely disappear. She did not attribute a cause to this change, which was not the subject of her investigation.

From the studies published in the 1960s, we turn to studies of more recent date. From her summary of palynostratigraphic data from Lower Cretaceous through Paleocene nonmarine formations in five regions along the Pacific coast of Russia, Markevich (1994) identified 14 palynostratigraphic zones. Of these, four that are in the Zeya–Bureya Basin pertain to the K–T boundary in the Russian Far East (Figure 8.12). Three are in the "Tsagayanskaya" or Tsagayan Formation (now Tsagayan Group), which is divided into lower, middle, and upper parts or "subformations" (now known as the Udurchukan, Bureya, and Darmakan formations, respectively), and one is in the "Kivdinskaya" or Kivda beds (now upper part of the Darmakan Formation); see Figure 8.10. Markevich's findings are summarized in Figure 8.13.

All similarities and differences considered, data presented by Markevich (1994) demonstrate strong parallels in the palynostratigraphy of the uppermost Cretaceous and lower Paleocene of the Russian Far East and western North America. The dilemma is in determining which aspects of similar patterns are most important in defining the K–T boundary. It appears that the K–T boundary is between either palynozones XI and XII or XII and XIII. In either case, it is evident that the stratigraphic ranges of some palynomorph species that occur in both eastern Asia and western North America are not identical, and it is clear that some pollen species that disappear at the K–T boundary in North America persist into the lower Paleocene in Russia.

Figure 8.10 encapsulates the significant differences in interpretation we have with our Russian colleagues about changes in palynofloras, dinosaur extinction,

Zone	Formation	Age	Taxa known in RFE and NA	Alternative Age
XIV	Kivda coal-bearing beds of Upper Tsagayan (upper Darmakan)	late Danian	*"Engelhardtioides"* (*Momipites*) *Pistillipollenites* *Polyvestibulopollenites* *Triatriopollenites* *Triporopollenites*	late Paleocene
XIII	Upper Tsagayan (lower Darmakan)	early Danian	*Liliacidites complexus* *Tricolpites microreticulatus* *Ulmoideipites krempii* *Wodehouseia fimbriata*	early Paleocene (with reworked Maastrichtian species?)
XII	Middle Tsagayan (Bureya)	late Maastrichtian	*Aquilapollenites quadrilobus* *Aquilapollenites spinulosus* *Orbiculapollis lucidus*	uncertain (includes both Maastrichtian and Paleocene species)
XI	Lower Tsagayan (Udurchukan)	early Maastrichtian	*Aquilapollenites conatus* *Aquilapollenites quadrilobus* *"Proteacidites"* (*Tschudypollis*) *thalmannii* *Wodehouseia spinata*	late Maastrichtian

Figure 8.13 Data on palynostratigraphic zones in the Zeya–Bureya Basin from Markevich (1994). Both the older and newer stratigraphic nomenclature (Figure 8.10) are shown. Palynomorph taxa listed are those that are known in both the Russian Far East (RFE) and North America (NA). Ages shown in the third column are from Markevich (1994); alternative ages shown in the fifth column are our interpretations based on occurrences of taxa in common between the RFE and NA. The K–T boundary may lie between either zones XI and XII or zones XII and XIII.

and the position of the K–T boundary. In this stratigraphic chart, the Upper Cretaceous–Paleocene and Maastrichtian–Danian (K–T) boundary is within the Tsagayan Formation or Group, at the contact between the Bureya and Darmakan formations (= Middle Tsagayan and Upper Tsagayan subformations). That position agrees with the zonation published by Markevich (1994) because it is at the contact between palynozones XII and XIII, as discussed above. The chart also shows, however, that the "main dinosaur horizon" is in the Udurchukan Formation (= Lower Tsagayan Subformation), with its top at a middle Maastrichtian–upper Maastrichtian boundary. Internationally, the Maastrichtian Stage is not formally subdivided, although Odin (1996) recommended dividing the Maastrichtian into two parts. In North America, the lower and upper parts of the Maastrichtian are informally recognized by many authors, but a middle Maastrichtian is not recognized. The apparent disappearance of the dinosaur fauna at the top of the "middle Maastrichtian" is a significant discrepancy and requires further investigation. If the position of the Maastrichtian–Danian boundary adopted by Russian stratigraphers (Figure 8.10) is to be accepted, it must be verified by the presence of iridium and shocked minerals, and supported by magnetostratigraphy and/or radiometric dating. None of these categories of supporting data are currently available in these sections.

Markevich *et al.* (2000) summarized evidence for vegetational change and dinosaur extinction in the Amur region and other localities in the Russian Far East, citing an abrupt change in climate and vegetation at the putative middle Maastrichtian–upper Maastrichtian boundary that resulted in the extinction of the dinosaurs. They presented palynological data to support the contention that climate change (cooling and increased humidity) occurred by the end of the mid Maastrichtian, as indicated by decreases in pollen of temperate plants and "unica" type pollen and increase in abundance of the Taxodiaceae. Markevich *et al.* concluded that dinosaur extinction was a global event that took place an estimated 2–3 million years before the end of the Cretaceous, in conjunction with the climatically induced change in vegetation. In support, they cited Nichols (1990), a paper mentioning that in the Lance and Hell Creek formations of Wyoming and Montana, dinosaur bones disappear 1–2 m below the K–T boundary. Presumably, Markevich *et al.* (2000) equated this small stratigraphic distance with their 2–3 million-year estimate, an equivalence that was neither implied by Nichols nor is supported by the data discussed here in Chapters 6 and 7. In fact, the 1–2 m interval represents only a tiny portion of the polarity subchron C29r and equates to between 10 000 and 100 000 years at the most. Markevich *et al.* reported that palynological assemblages similar to those in the Amur region are also present in the Chukotka region of Siberia and on Sakhalin Island, and that all of them are "middle Maastrichtian" in age.

Markevich and Bugdaeva (2001) reiterated the assertion that both the vegetational change and dinosaur extinction in the Russian Far East took place at the locally defined middle Maastrichtian–upper Maastrichtian boundary. They asserted that the tripartite division of the Maastrichtian Stage is based on paleontological data from marine invertebrates (inoceramids) and terrestrial vertebrates (dinosaurs), as well as from paleobotany (Golovneva 1994a, b) and palynology (Markevich *et al.* 1994).

A palynostratigraphic summary for Kundur (locality 97) in the Amur region published by Markevich and Bugdaeva (2001) and Bugdaeva (2001) (Figure 8.14)

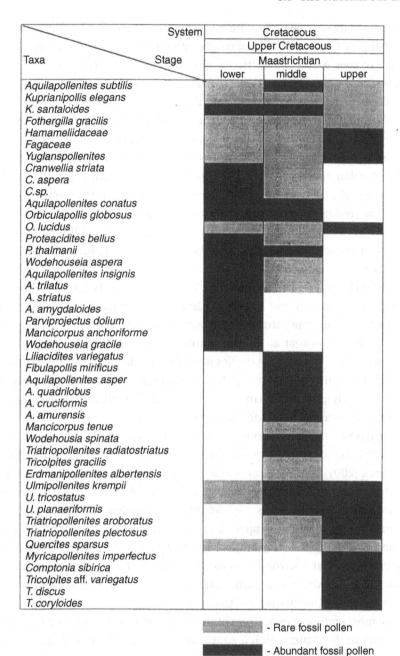

Figure 8.14 Palynostratigraphy of the Kundur locality, Russian Far East (from Markevich and Bugdaeva 2001). Reprinted by permission.

shows a strong change in the composition of the angiosperm pollen assemblage occurring at their "middle Maastrichtian–upper Maastrichtian boundary." In the list of species, we call particular attention to *Aquilapollenites conatus*, *A. quadrilobus*, *Cranwellia striata* (= *C. rumseyensis*), *Orbiculapollis lucidus*, *Proteacidites thalmannii* (= *Tschudypollis thalmannii*), and *Wodehouseia spinata*. All are species of the *Wodehouseia spinata* Assemblage Zone of the United States and Canada, which is late (not "middle") Maastrichtian in age (Nichols and Sweet 1993, Nichols 1994). The abrupt disappearance of these species (*O. lucidus* excepted) along with 16 other Russian species (55% of those listed) creates a pattern like that at the palynological K–T boundary in western North America. We would argue that those palynological extinctions at the Kundur locality of Bugdaeva (2001) and Markevich and Bugdaeva (2001) mark, not the middle Maastrichtian–upper Maastrichtian boundary, but the upper Maastrichtian–Paleocene (K–T) boundary.

A palynostratigraphic summary for Sinegorsk (locality 99) on Sakhalin presented by Markevich and Bugdaeva (2001) and reprinted in Bugdaeva (2001) shows much the same pattern (Figure 8.15). The angiosperm pollen assemblage there is similar to that at Kundur, although not identical in composition. At Sinegorsk, three of the same species named above disappear at the same level as 44% of the individual species listed. Here again, we would reason that the profound palynological change marks the K–T boundary, not the middle Maastrichtian–upper Maastrichtian boundary.

A palynostratigraphic summary for localities such as Beringovskoe (locality 98) in the Koryak Upland prepared by Markevich and Bugdaeva (2001) and Bugdaeva (2001) lists the same angiosperm taxa as for Kundur and Sinegorsk, but the pattern of occurrences is less clear (Figure 8.16). Only two of the species named above, *Aquilapollenites quadrilobus* and *Proteacidites thalmannii*, disappear at the middle Maastrichtian–upper Maastrichtian boundary; the other four either do not extend to that stratigraphic level or they transcend it. The extinction percentage at that horizon drops to 23%. Still, the change in assemblages may mark the K–T boundary in northeastern Siberia, but it is less clear than in the other regions of the Russian Far East.

The most complete exposition of the flora and dinosaurs at and near the K–T boundary in the Amur region is contained in a book edited by Bugdaeva (2001). In addition to the three range charts reproduced here as Figures 8.14 through 8.16, the book thoroughly documents the stratigraphy, flora, and fauna of the Zeya–Bureya Basin. A stratigraphic chart in the Bugdaeva (2001) book shows the palynozones of Markevich (1994) discussed here previously (Figure 8.12 and Figure 8.13). A difference is that zone XI, which Markevich had earlier maintained was in the lower Maastrichtian, is shown as both lower and middle

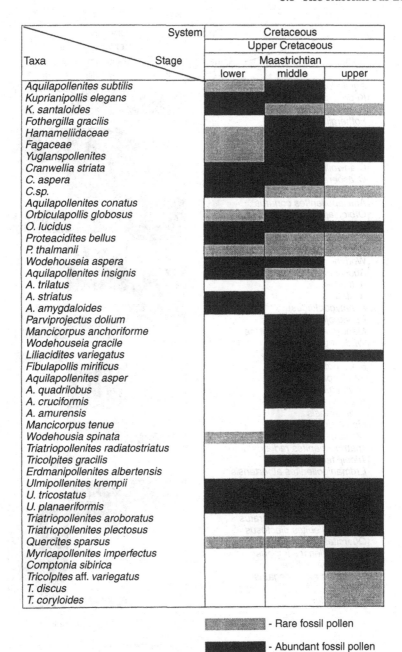

Figure 8.15 Palynostratigraphy of the Sinegorsk locality, Russian Far East (from Markevich and Bugdaeva 2001). Reprinted by permission.

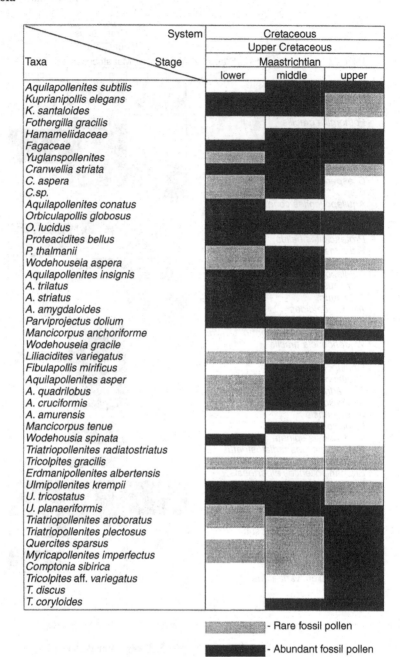

Figure 8.16 Palynostratigraphy of the Beringovskoe locality, Russian Far East (from Markevich and Bugdaeva 2001). Reprinted by permission.

Stratigraphic interval	Localities	Palynomorph assemblages
Danian (Zones XIII–XIV)	Tsagayan Fm. stratotype at Belaya Gora	Pollen of the Betulaceae, Myricaceae, and Juglandaceae dominates; Fagaceae, Salicaceae, and Ulmaceae pollen present; *Aquilapollenites spinulosus* and *A. subtilis* present in lower part of section, but *Aquilapollenites* and *Wodehouseia* disappear in upper part; *Caryapollenites* present
Upper Maastrichtian (Zone XII)	Mutnaya and Udurchukan River area and Belaya Gora	New species of *Aquilapollenites* appear, but *Aquilapollenites* decreases in abundance; species of *Triatriopollenites* are characteristic; *Orbiculapollis* and *Wodehouseia* increase in abundance; *Anacolosidites* and *Kurtzipites* are present
Middle Maastrichtian (Zone XI)	Blagoveshchensk area and stratotype of Tsagayan Fm., Russia, and Yuliangzi Fm., China	Angiosperm pollen dominates all assemblages; gymnosperms also abundant; fern and moss spores diverse; ten species of *Aquilapollenites* present; *Orbiculapollis lucidus*, "*Proteacidites*" (*Tschudypollis*) and *Wodehouseia spinata* present
Lower Maastrichtian (Zone XI)	Kundur area, Zeya–Bureya Basin	Fern spores and gymnosperm pollen of Taxodiaceae and Gnetaceae common; angiosperm pollen (*Triatriopollenites, Tricolpites, Tricoloporopollenites,* and *Triporopollenites*); *Aquilapollenites* and other triprojectates nominally present

Figure 8.17 A summary of palynostratigraphic data from Bugdaeva (2001). Note that Zone XI encompasses both the lower and middle Maastrichtian, and that the Danian, which is considered to be the entire Paleocene, includes Zone XIII in its lower part and Zone XIV in its upper part. *Caryapollenites*, which is reported from the lower Danian in the Zeya–Bureya Basin, is indicative of the upper Paleocene in North America.

Maastrichtian; zone XII is still shown as upper Maastrichtian. Palynologic data from Bugdaeva (2001) are summarized here in Figure 8.17.

In a discussion of the Maastrichtian Stage and correlation of plant-bearing deposits of the Russian Far East in Bugdaeva (2001), the palynofloras of the Lower Tsagayan Subformation (= Udurchukan Formation) are assigned to palynozone XI (*Wodehouseia spinata–Aquilapollenites subtilis* palynozone). Discoveries of dinosaur bones in the Lower Tsagayan and in the Yuliangzi Formation across the Amur River in China are reported and the dinosaur-bearing beds of the Amur region are palynologically correlated with their stratigraphic equivalent, the Yuliangzi Formation in Heilongjiang Province. A sharp reduction in floral diversity associated with the disappearance of the dinosaurs in the Amur and Heilongjiang regions is reported at the middle–upper Maastrichtian boundary. However, we would equate both events – floral change and dinosaur

Figure 8.18 View of the highwall at the Arkhara–Boguchan coal mine showing the upper part of the Middle Tsagayan and lower part of the Upper Tsagayan subformations.

extinction – with the Maastrichtian–Paleocene (K–T) boundary, as is the case in North America.

We have some data of our own to contribute to the debate about the ages of the lower, middle, and upper parts of the Tsagayan Group. One of us (DJN) collected palynological samples at the localities discussed by Markevich and Bugdaeva in Bugdaeva (2001). The localities include outcrops near the town of Kundur and coal mines near the towns of Raichikhinsk and Arkhara. Samples from the Lower Tsagayan Subformation (= Udurchukan Formation) near the town of Kundur yielded palynomorphs indicative of Maastrichtian age, which is partly consistent with the rather broad age range for the Tsagayan Group shown in Figure 8.10. As anticipated, samples from the Kivda beds of the Upper Tsagayan Subformation (= upper, coal-bearing part of the Darmakan Formation) in coal mines near Raichikhinsk yielded Paleocene palynomorphs. Our samples from the Arkhara–Boguchan coal mine (Figure 8.18) were the most interesting. They were from the upper part of the Middle Tsagayan and the overlying lower part of the Upper Tsagayan subformations (= Bureya and Darmakan formations of Figure 8.10) and were said by Russian geologists possibly to span the K–T boundary. However, sixteen samples collected through 24 m of exposure in the highwall of the mine yielded only Paleocene palynomorphs (Figure 8.19), including species of *Momipites* well known from numerous localities in western North America (Nichols 2003). Clearly, this

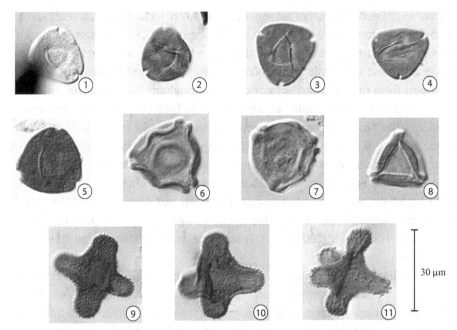

Figure 8.19 Fossil pollen of Paleocene age from the upper part of the Bureya Formation and lower part of the Darmakan Formation at the Arkhara–Boguchan coal mine; identifications based on closely similar or identical North American species. 1–4 - *Momipites tenuipolus*, 5 - *Momipites* sp. cf. *M. wyomingensis*, 6 - *Paraalnipollenites confusus*, 7 - *Triporopollenites infrequens*, 8 - "*Paliurus*" *triplicatus*, 9–11 - *Aquilapollenites spinulosus* (= *A. subtilis*).

section does not include the K–T boundary. These results suggest that the K–T boundary lies below the sampled interval, near the boundary between the "middle" and upper Maastrichtian of Figure 8.10. They are also in general agreement with those of Markevich *et al.* (2004) and in complete agreement with the current interpretation of Valentina Markevich (personal communication, 2007).

There might appear to be a remaining obstacle with our reinterpreting the age of the Udurchukan Formation and the "main dinosaur horizon" (Figure 8.10) in the Amur region as late Maastrichtian: the age of the dinosaur fauna associated with the palynomorph assemblages, but there is evidence in support of our opinion. Godefroit *et al.* (2003) concluded that a dinosaur from the lower part of the Tsagayan Formation (= Udurchukan Formation) at the Kundur locality was either "middle" Maastrichtian, in accordance with the Russian interpretations discussed previously, or – more likely – was late Maastrichtian in age. The dinosaur was found in association with palynomorphs that Godefroit *et al.* recognized as belonging to zone XI of Markevich (1994), the *Wodehouseia spinata–Aquilapollenites subtilis* palynozone. Van Itterbeeck *et al.* (2005) reached the same conclusion about the palynological age of the Udurchukan Formation being late Maastrichtian. Additionally, from a locality just across the Amur River in China, Godefroit *et al.* (2001) had previously determined the age of a dinosaur from the Yuliangzi Formation in Heilongjiang Province. The Yuliangzi Formation is the stratigraphic equivalent of the Udurchukan Formation and it contains the "main dinosaur horizon" in that part of China (see Figure 8.9). Godefroit *et al.* (2001) identified the palynological assemblage from the Yuliangzi Formation as the *Wodehouseia spinata–Aquilapollenites subtilis* palyno-zone, and they equated it with assemblages from formations of "Lancian" (late Maastrichtian) age in North America. Based on our knowledge of the North American palynofloras of late Maastrichtian age, we are inclined to concur with Godefroit *et al.*'s (2001, 2003) interpretations and that of Van Itterbeeck *et al.* (2005).

Our colleagues Valentina Markevich and Eugenia Bugdaeva continue to assert (personal communication, 2007) that a palynozone of latest Maastrichtian age lies above the stratigraphic level from which Godefroit *et al.* (2003) collected the dinosaur. According to Markevich, this palynozone is char-acterized by distinctive species of *Aquilapollenites* and related triprojectate pollen that are not found in association with the dinosaurs of the Udurchukan Formation. For us, this debate remains unresolved because the key pollen species are endemic to the Russian Far East and unknown in western North America, and the strata of the overlying Bureya Formation have not yielded stratigraphically useful plant megafossils. Furthermore, it must be recognized that independent age indicators (paleontologic, geochronologic, or magneto-stratigraphic) have yet to be found in the Maastrichtian strata of the Russian Far East. At best, the resolution of the K–T boundary in this region is at the stage level (see Section 2.1).

The Cretaceous and Paleocene megafloral record from the Russian Far East is extensive, and numerous Russian paleobotanists have expressed strong opi-nions concerning the nature of megafloral response to the K–T boundary. They are most recently summarized by Golovneva (1995), Herman and Spicer (1995, 1997), Krassilov (2003), and Akhmetiev (2004). All of these authors have argued against a catastrophic megafloral extinction and have posited climate or environmental change as causal factors. It is our opinion that these conclusions are somewhat premature because the K–T boundary itself has not been inde-pendently located at any site in the Russian Far East. Thus, many of the argu-ments for floral change across the boundary appear to be somewhat circular. We have already seen that there is a difference in the Asian and North American interpretations of the palynological K–T boundary, and for the moment, the interpretation of the megafloral record hangs on that of the palynofloral record. Resolution of this situation awaits the development of a reliable geochronology and magnetostratigraphy for the region and/or the discovery and documentation of the K–T boundary impactite itself.

Three primary areas have supplied the majority of the data for the discussion of the K–T boundary in the Russian Far East. These are the exposures along the forested north bank of the Amur River (Krassilov 1976, Kodrul 2004), a series of exposures in the tundra of the Koryak Uplands (Golovneva 1994a, 1994b, 1995, Herman and Spicer 1995, 1997), and beach cliff exposures on the western margin of Sakhalin Island (Krassilov 1978, 1979, 2003).

The situation in the Amur is basically this: megafloras occur in association with dinosaurs in the Kundur Formation, they are essentially absent in the Lower Tsagayan Subformation (= Udurchukan Formation), and they occur with Paleocene palynofloras in the Middle and Upper Tsagayan subformations (= Bureya and Darmakan formations); Figure 8.10. This is essentially the same situation seen on the south bank of the Amur River in China, where the Taipinglinchang and dinosaur-bearing Yuliangze formations are separated from the leaf- and pollen-bearing Wuyun Formation by the poorly dated and poorly sampled Furao Formation (Figure 8.9). The Cretaceous megafloras from both the Kundur and Taipinglinchang formations are only preliminarily known (Golovneva *et al.* 2004), whereas the unambiguously Paleocene floras from the Wuyun and Upper Tsagayan (Darmakan) formations are well described (Krassilov 1976, Kodrul 2004). This is somewhat similar to the situation in the USA in the late 1980s before the flora of the upper Maastrichtian Hell Creek Formation was first described. The Paleocene flora was known, but the Cretaceous less so. In addition, the actual ages of the uppermost Cretaceous units in the Amur region are known only to the stage level. At best, this can be described as stage-level temporal resolution, and further understanding of these sections awaits the collection of more fossils and independent dating of the Furao Formation on the Chinese side and the Middle Tsagayan Subformation (Bureya Formation) on the Russian side.

In the Koryak Uplands, Golovneva (1994a, b) collected 124 megafloral localities from the 900–2000-m-thick Rarytkin Formation exposed along Rarytkin Ridge, a northeast trending range that is cut by the Anadyr River. From this formation, she described the "middle Maastrichtian" Gornorechensk floral assemblage (Gornorechenian floral stage) and the "late Maastrichtian–Danian" Rarytkin floral assemblage (Rarytkinian floral stage). In these areas, it was not possible to measure stratigraphic sections and the relative stratigraphic positions of the megafloral localities were extrapolated (Golovneva, personal communication, 2000). The Cretaceous age of the lower part of the section is constrained by marine mollusks and dinosaurs known from correlative beds in the Kakanut River Basin some 160 km to the south. Golovneva defined the position of the K–T boundary in Rarytkin Ridge by correlation of Cretaceous marine mollusks associated with Rarytkinian floral-stage fossils in the Khatyrka River Basin more than 180 km to the southwest. Based on these lengthy correlations, she interpreted the significant floral change at the Gornorechenian–Rarytkinian floral-stage boundary to occur at the middle Maastrichtian–upper Maastrichtian boundary. Based on this interpretation, Golovneva argued that floral change at the K–T boundary itself was insignificant. We would argue that the age control on this section is poor and that statements about megafloral change at the K–T boundary in this area are premature. Subsequent work by Herman and Spicer (1997) also used correlation of Cretaceous mollusks to argue for a Cretaceous age for the Rarytkinian floral-stage megaflora; they provocatively titled their paper with a temporal tautology, "The Koryak flora: did the early Tertiary deciduous flora begin in the late Maastrichtian of northeastern Russia?" This is not the first time that the Arctic has been implicated with the premature appearance of a floral stage, but as with past cases, the burden of proof is with the proposal, and independent age assessment is desperately needed before the relevance of these sections to the K–T boundary can be established.

On Sakhalin Island, Krassilov (1978, 1979, 2003) described the "Maastrichtian" Augustovian megaflora from the upper part of the Krasnoyarka Formation in sections along the Augustovka River and an overlying "Maastrichtian–Danian" Boshnyakovian flora from the Boshnyakova Formation. In these sections as well, direct age control remains a significant concern and assertions about floral change at the K–T boundary should be treated as preliminary.

In general, it appears that the Russian paleobotanists have been too quick to assign global stage names and associated temporal significance to local and poorly dated terrestrial sections in the Russian Far East. These sections have clear potential, but to date it has not been realized, and it is premature to discuss the significance of these floras as if they were known to chron- or impactite-level resolution.

9

The remnants of Gondwana

9.1 Overview

From the palynomorph- and leaf-bearing intervals of latest Cretaceous and early Paleocene age in North America and Eurasia, we now direct our quest for records of plants and the K–T boundary to lands formerly or currently in the equatorial region or the Southern Hemisphere. These lands are the remnants of Gondwana. The continents formed by the break-up of Gondwana were well separated by K–T boundary time. Africa, India, and South America had moved into tropical latitudes while Australia, New Zealand, Madagascar, and Antarctica remained at mid to high latitudes.

Herngreen *et al.* (1996) stated that, in Late Cretaceous time, South America and most of Africa lay within the Palmae palynofloral province, which is characterized by assemblages with 10–50% pollen of the type produced by the Arecaceae (palms) and related species. That palms characterized the floras of the equatorial regions of these continents even in the Late Cretaceous is not surprising. Northern Africa was well north of the equator at 66 Ma and close to Europe and the Normapolles Province. As a result, northern Africa was in a transition zone between the Palmae and Normapolles provinces (Herngreen *et al.* 1996); see Figure 5.4. There was an increase in Normapolles pollen in northern Africa in the early Paleocene. Contemporaneous palynofloral assemblages from India included elements of the Palmae Province admixed with some from both the Normapolles and *Aquilapollenites* provinces; there is no obvious paleogeographic explanation for this. Axelrod and Raven (1978) and Maley (1996) asserted that rainforests existed in Africa in Late Cretaceous time. Strong changes in palynofloras across the K–T boundary are not recorded for any of the equatorial regions, a pattern that led Hickey (1981) to argue that floral extinction was not global in extent. From our point of view, there has yet to be a well-studied and temporally constrained K–T boundary in the paleotropics, so this interpretation is premature.

With the notable exception of New Zealand, the Southern Hemisphere has yielded very little data on the fossil record of plants across the K–T boundary. There is broad-scale palynostratigraphic data from Australia, but nothing that directly pertains to the K–T boundary, other than that it coincides with the boundary between two palynological biozones in rocks not exposed at the surface. There is some indirect information from Antarctica, discussed below. There is little known from South America, India, Madagascar, and Africa. The meager record of the K–T boundary is largely due to the apparent absence of nonmarine rocks of latest Cretaceous and early Paleocene age, although more may be learned with future exploration.

9.2   South America

In latest Cretaceous time, northern South America lay within the Palmae palynologic province (Herngreen and Chlonova 1981). Characteristic pollen included species of the genera *Psilamonocolpites*, *Retimonocolpites*, *Proxapertites*, and *Spinizonocolpites* (the species *S. echinatus* is said to be identical with pollen of the living palm *Nypa fruticans*). Muller *et al.* (1987) presented a comprehensive palynostratigraphy of the Cretaceous through Holocene of northern South America. Their study areas included Colombia, Venezuela, Trinidad, Guyana, Surinam, and Brazil. Much of their information came from published sources, especially from Germeraad *et al.* (1968). Zone 13 of Muller *et al.* (Maastrichtian) is defined by the lowest stratigraphic occurrence of *Proteacidites dehaani*, which becomes extinct at the top of the zone, along with *Buttina andreevi*, *Crassitricolporites brasiliensis*, species of *Aquilapollenites*, and *Scollardia* (the last two are genera also known in the Maastrichtian of North America). Zone 14 (basal Paleocene) is defined by lowest stratigraphic occurrence of *Spinizonocolpites baculatus* (which is palm pollen). Hence, a K–T boundary can be recognized by a change in palynofloras, but apparently not a major extinction event. The lowermost Paleocene is characterized by low abundance of fern spores and gymnosperm pollen, and by having first occurrences of pollen of the largely tropical family Bombacaceae and the palm genus *Mauritia*, both of which are dominant in northern South America at present. Specimens of some of these genera and species are illustrated in Figures 9.1 and 9.2.

Ashraf and Stinnesbeck (1988) described the palynology of a K–T boundary interval near Recife, on the coast of eastern Brazil. Foraminifera were used to determine the age of the section. Ashraf and Stinnesbeck reported that the

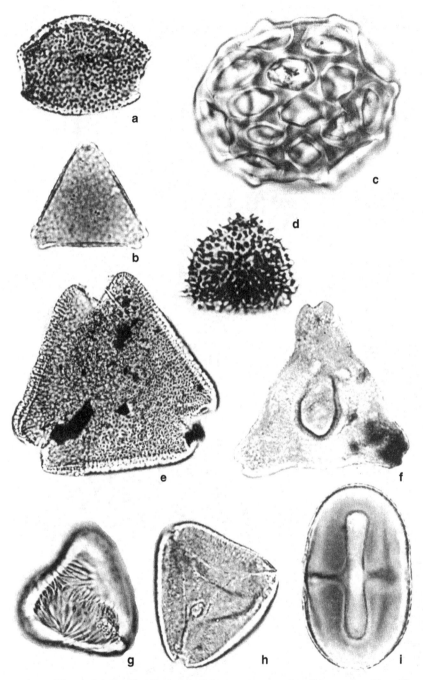

Figure 9.1 Fossil pollen of the Palmae palynostratigraphic province (from Herngreen and Chlonova 1981). a – *Retidiporites magdalenensis*, b – *Proteacidites sigalii*, c – *Buttinia andreevi*, d – *Echitriporites trianguliformis*, e – *Foveotricolpites irregularis*, f – *Aquilapollenites sergipensis*, g – *Scollardia srivastavae*, h – *Cupanieidites* sp., i – *Crassitricolporites brasiliensis*. Specimens range from 30 to 60 micrometers in diameter.

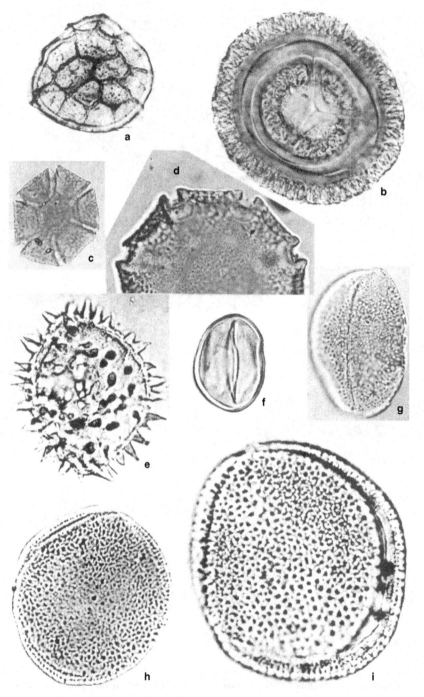

Figure 9.2 Fossil pollen of the Palmae palynostratigraphic province (from Herngreen and Chlonova 1981). a – *Zlivisporis* (?*Triporoletes*) *blanensis*, b – *Gabonisporis vigourouxii*, c – *Psilastephanocolporites daportae*, d – *Psilastephanocolporites hoekenae*, e – *Spinizonocolpites echinatus*, f – *Psilamonocolpites medius*, g – *Retimonocolpites* sp., h – *Proxapertites operculatus*, i –*Proxapertites cursus*. Specimens range from 30 to 70 micrometers in diameter.

palynoflora is composed primarily of fern spores and made no mention of the pollen species discussed by Muller *et al.* (1987). There is no palynologic break evident at the position of K–T boundary established by Ashraf and Stinnesbeck. They concluded that the palynoflora shows climate change from subtropical in the Maastrichtian to subtropical to temperate in the Paleocene. Given the absence of characteristic angiosperm pollen, this record is ambiguous at best.

With particular reference to western Venezuela, Pocknall *et al.* (2001) stated that the transition from the Maastrichtian to the Paleocene in northern South America usually occurs at a change from predominantly marine depositional conditions of the uppermost Cretaceous to continental (fluvial and lacustrine) conditions in the lowermost Paleocene. There may be some uncertainty in the dating of this transitional interval, however, because it is based solely on terrestrially derived spores and pollen. Correlation with the marine fossil record has not been possible. A measured section through the Mito Juan Formation (Maastrichtian) and the Los Cuervos Formation (Paleocene) in the Andes of western Venezuela yielded diverse palynomorph assemblages that include dino-flagellates. Graphic correlation of the Venezuelan dinoflagellate data and a dinoflagellate-biostratigraphic database indicates the presence of three uncon-formities within this interval. Nonetheless, Pocknall *et al.* stated that a conform-able K–T boundary section might be preserved within a covered interval about 20 m thick, assuming continuous deposition and preservation of sediments. The dinoflagellate taxa present in the measured section indicate that the strata below the covered interval are latest Cretaceous in age and that those just above are earliest Paleocene in age. The data available at present leave the precise position of the K–T boundary unresolved, and associated changes in terrestrial vegetation unclear.

Correlation from the measured section in western Venezuela studied by Pocknall *et al.* (2001) to their global database containing marine dinoflagellates helps to refine the age ranges of the pollen species that have been used to define the Maastrichtian *Proteacidites dehaani* and Paleocene *Spinizonocolpites baculatus* pollen zones in South America (e.g., Muller *et al.* 1987). The fungal fruiting body *Trichopeltinites*, which is known to disappear at or about the K–T boundary in sections in North America (see Section 7.2), is also recorded in western Venezuela, although in Venezuela it apparently ranges into the Paleocene. Thus, the most recent data available confirm earlier conclusions, some of which date back to 1968.

Recent fieldwork in Chubut Province, Argentina, has begun to elucidate Late Cretaceous and early Paleocene palynofloras and megafloras, but this work is largely unpublished and stratigraphic resolution is currently at the stage level. In late 2006, fieldwork by one of us (KRJ) in concert with Museo Paleontológico

Egidio Feruglio in Trelew and Pennsylvania State University, located several plant-bearing horizons in the primarily marine Maastrichtian–Paleocene Lefipan Formation in the upper Chubut River valley. We made large collections from several angiosperm-dominated floras and collected ancillary foraminiferal, palynological, and magnetostratigraphic samples in order to acquire age control. Because Maastrichtian or Paleocene megafloras are very poorly known in South America, plant megafossils alone are not sufficient to determine whether these sites are Cretaceous or Paleocene.

In 2004–2006, La Plata University student Ari Iglesias collected several megafloral localities in southern Chubut near the town of Sarmiento. He found several leaf localities at and near a well-known petrified forest in the Salamanca Formation. In this area, this formation contains distinctive Danian marine foraminifera, and additional paleomagnetic data constrains this megaflora to ~62 Ma. The Palacio de los Loros flora is considerably richer than coeval megafloras from North America; a collection of more than 2500 specimens yielded an angiosperm-dominated flora of 36 species (Iglesias *et al.* 2006, 2007).

Perhaps the most promising area for K–T boundary studies in South America is Colombia. Carlos Jaramillo and his student Felipe De la Parra have recently recovered a 500 m core from the Cesar–Rancheria Basin. Preliminary analysis of 62 samples and more than 14 000 palynomorph specimens suggests a K–T boundary extinction on the order of 60–70% (Jaramillo and De la Parra 2006). This work has not yet been calibrated with paleomagnetics, so the precise age relations are not yet clear. Nonetheless, the proximity to Chicxulub and the apparently high extinction suggest that this region holds much promise.

9.3   Africa

In their paper on global palynofloristic provinces, Herngreen and Chlonova (1981) listed the characteristic taxa of the Palmae Province, which existed in Africa and northern South America in Senonian (Coniacian through Maastrichtian) time. Most noteworthy among them are the palm pollen genera *Psilamonocolpites*, *Retimonocolpites*, *Proxapertites*, and *Spinizonocolpites*. Assemblages in which palm pollen is prominent characterize upper Senonian strata, but angiosperm pollen becomes increasingly diverse. Among the angiosperms are the morphologically distinctive *Buttinia andreevi* and *Droseridites senonicus*; the syncolporate genus *Cupanieidites*; and species of the triporate genera *Echitriporites*, *Proteacidites*, and *Triorites* (especially *E. trianguliformis*, *P. dehaani*, and *P. sigalii*). Rare occurrences of species more typical of the Normapolles and *Aquilapollenites* provinces are also noted within the Palmae Province in Africa, and some African genera are also known in contemporaneous rocks in India (Herngreen and Chlonova 1981).

Kaska (1989) summarized the Lower Cretaceous to Tertiary palynostratigraphy of Sudan. A composite stratigraphic section ranging in age from Barremian through Paleogene included only five palynostratigraphic zones. Zone C, which encompasses most of the Upper Cretaceous, ranges from Turonian through Maastrichtian; it is overlain by Zone D of Paleocene–Eocene age. The palynoflora of the uppermost Cretaceous includes many of the key species of the Palmae Province mentioned by Herngreen and Chlonova (1981). Kaska (1989) noted that the common markers for the uppermost Cretaceous are *Proteacidites sigalii*, *Buttinia andreevi*, and *Echitriporites trianguliformis*, and that rare species of *Aquilapollenites* and Normapolles pollen are also present in the Sudan section. Kaska stated that, in his composite section, there is a transitional interval between Zone C and the overlying Zone D that is at least in part Paleocene in age. The interval is composed largely of lacustrine beds, in which the most characteristic species is the freshwater fern megaspore *Ariadnaesporites*. Clearly, no K–T boundary can be recognized in Sudan at this time.

Salard-Cheboldaeff (1990) compiled palynostratigraphic data from the Maastrichtian to the Quaternary of western Africa, in Senegal, Ivory Coast, Ghana, Benin, Gabon, Congo, and Angola. From these data, she concluded that the palynological assemblages from this intertropical area differ significantly from those of northern Africa but have much in common with those of South America and to a lesser extent with those of India. Salard-Cheboldaeff's species list from Benin includes *Buttinia andreevi*, *Proteacidites dehaani*, *P. sigalii*, and a species of *Spinizonocolpites* in the upper Maastrichtian. In Cameroon, *Proteacidites dehaani* and *Spinizonocolpites*, among others, are present in the upper Maastrichtian, and a species of *Proxapertites* characterizes the Paleocene. Thus, the pattern appears to be familiar, with typical species of the Palmae Province present in the uppermost Cretaceous of western Africa. Palms such as those represented by *Proxapertites* pollen continued to develop in the Paleocene while gymnosperms declined. The pattern of stratigraphic ranges of 163 mostly species-level taxa plotted by Salard-Cheboldaeff (1990) showed no significant extinction across the K–T boundary.

Salard-Cheboldaeff (1990) observed that palynological assemblages from the K–T boundary interval in western and central Africa differ from those of northern Africa. The most extensive data from northern Africa are those of Méon (1990, 1991). Méon studied the terrestrial pollen and spores in the well-known marine K–T boundary section at El Kef, Tunisia (locality 100 in Table 2.1 and Appendix). The position of the boundary had been determined by foraminiferal biostratigraphy. Méon tracked occurrences of 56 spore and 157 pollen taxa within the interval at El Kef. In the uppermost Maastrichtian, spores and pollen of the gymnosperm family Araucariaceae are most abundant, and palm pollen is present. The palm pollen shows the association of this palynoflora with the

Palmae Province, but pollen of the Normapolles Group is present, as well. The latter species have a European affinity and serve to distinguish this palynoflora from those of western and central Africa, as noted by Salard-Cheboldaeff (1990). In the Paleocene part of the section, palm pollen increases in abundance along with pollen of the Normapolles Group and the family Juglandaceae. Thus, the Paleocene assemblages of northern Africa show more affinity with those of Europe than with those of western and central Africa or South America.

Méon's (1990, 1991) study of the K–T boundary interval at El Kef presents intriguing data. The palynoflora became impoverished in the upper Maastrichtian, but Méon ruled out a catastrophic cause for changes in palynological assemblages at the K–T boundary. She interpreted the change across the boundary to be the result of a gradual, climatically induced extinction. Other interpretations are possible, however. Méon (1990) reported that 55 palynomorph species disappear before or at the boundary. Méon's plot of stratigraphic range data (Figure 9.3) shows about 250 m of section in which disappearances proceed at a steady rate up to the K–T boundary (Figure 9.4). From Méon's data we calculate that 13.6% of palynomorph taxa occurring in the uppermost zone of the Cretaceous (the unnamed zone just above the *Abatomphalus mayorensis* foraminiferal zone) disappear at the K–T boundary, either in the sample at the boundary or the one just below it. Song and Huang (1997), citing the data of Méon (1991), reported 53.3–55% extinction across the K–T boundary at El Kef, although we are unsure how they arrived at such a high number. Regardless of the actual percentage, the pattern of range truncations of palynomorph taxa at El Kef shows a significant increase in the extinction rate of terrestrial palynomorphs over background extinctions at the K–T boundary.

The record from El Kef, Tunisia, is especially noteworthy because the International Commission on Stratigraphy recognizes it as the Global Stratotype Section and Point (GSSP) that marks the base of the Paleogene System, thus it can be regarded as the type section of the K–T boundary. The GSSP at El Kef is characterized by an iridium anomaly and foraminiferal extinctions. For our purposes in trying to determine the nature of the K–T boundary in the terrestrial realm, however, the marine section at El Kef is less than ideal. An extinction level among palynomorphs of terrestrial origin ranging from 13% (our figure) to 55% (Song and Huang 1997) must be regarded as significant because percentage extinction at nonmarine boundary sections in North America is on the order of 15–30% (see Section 6.3).

9.4   India

The position of the K–T boundary in India is of special interest because many authors have suggested the Deccan Traps, one of the largest terrestrial

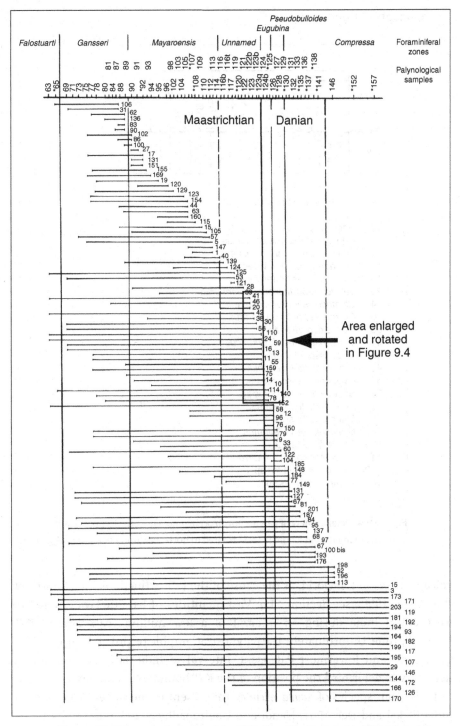

Figure 9.3 Palynostratigraphy of the El Kef locality, Tunisia (modified from Méon 1990). The Maastrichtian–Danian (K–T) boundary is between an unnamed foraminiferal zone ("non nommee") and the *Globigerina eugubina* foraminiferal zone ("*Eugubina*"). The area within the box showing palynomorph occurrences near the boundary is enlarged and rotated in Figure 9.4. Reprinted by permission of Elsevier.

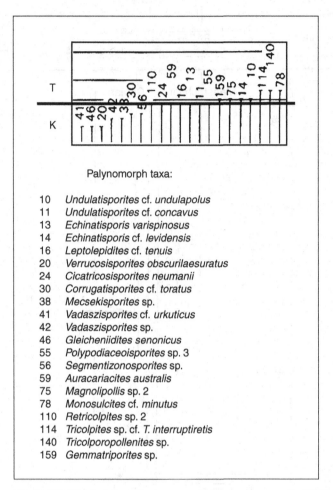

Figure 9.4 Detailed view of palynomorph occurrences near the K–T boundary at El Kef, Tunisia, enlarged from Figure 9.3 and rotated to horizontal (modified from Méon 1990).

volcanic flows in the world, as the cause of the K–T extinctions (e.g., Courtillot *et al.* 1988, 1990); see Section 11.4. Vast quantities of basaltic lava erupted in west-central India during an interval of a few million years around K–T boundary time. Unfortunately, the palynostratigraphic record of the uppermost Cretaceous and lowermost Paleogene interval in India lacks the resolution necessary to determine the position of the K–T boundary accurately.

Indigenous species of *Aquilapollenites* are present in the Upper Cretaceous in India (Baksi and Deb 1981, Nandi and Chattopadhyay 2002), but India is not considered to be within the *Aquilapollenites* Province. Herngreen *et al.* (1996) assigned India to the Palmae Province (see Figure 5.4). Paleocene assemblages show affinity with both the Palmae and Normapolles provinces, as discussed by

Herngreen *et al.* (1996). Based on their study of a "subcrop sequence" in the Bengal Basin, Baksi and Deb (1981) stated that the K–T boundary is within the Jalangi Formation, but they published no details. Nandi (1990) reported on Upper Cretaceous palynostratigraphy of northeastern India, but the Senonian assemblages are imprecisely dated and of little or no value in determining the position of the K–T boundary. Dogra *et al.* (1994) conducted a palynostratigraphic study of the Lameta Formation in Madhya Pradesh, central India, determining that it is Maastrichtian in age, but there is no discussion of the K–T boundary in that report.

Alt *et al.* (1988) suggested that an extraterrestrial impact might have initiated the extensive flows of basaltic lavas known as the Deccan Traps, which might have linked the two mechanisms as the cause of the K–T extinctions. This is unlikely, however, because the Deccan Trap flows are dated as having erupted between 67 and 63 Ma (Venkatesan *et al.* 1993, 1996) and thus are only partly coincident with the K–T impact event at 65.5 Ma. Within the intertrappean sequence, Bhandari *et al.* (1995, 1996) discovered an iridium anomaly between Flow III and Flow IV at Anjar, Kutch district, Gujarat, western India. The level is dated at 65.5 ± 0.7 and 65.4 ± 0.7 Ma. Bhandari *et al.* (1995) concluded that the impact event was the origin of the iridium and could not have triggered the Deccan volcanism because the lava flows had begun about a million and a half years earlier. On the basis of magnetic susceptibility and carbon isotope data, Hansen *et al.* (2001) disputed Bhandari *et al.*'s (1995) claim that their section was at the K–T boundary, but the Hansen *et al.* paper was contradicted by Shukla and Shukla (2002), who supported Bhandari *et al.*'s assessment on geochemical and geochronological grounds. Courtillot *et al.* (2000) also provided supporting data for the Anjar section.

Dogra *et al.* (2004) investigated the palynology of the locality studied by Bhandari *et al.* (1995) at Anjar, western India (locality 101 in Table 2.1 and Appendix). Dogra *et al.* determined that their assemblage, which included *Aquilapollenites indicus* and species assigned to the genus *Proteacidites*, is Maastrichtian in age. Singh *et al.* (2006) found that a palynomorph assemblage from mammal-bearing intertrappean beds in Andra Pradesh, southeastern India, is Maastrichtian in age. To date, no palynomorphs of Paleocene age have been recovered from the intertrappean sedimentary strata.

9.5   Antarctica

The paleobotanical and palynological record of the K–T boundary at the southern end of the world – in Antarctica – is confounded by the rarity of uppermost Cretaceous and lower Paleocene rocks deposited in nonmarine environments. There is no megafossil paleobotanical record that spans the K–T

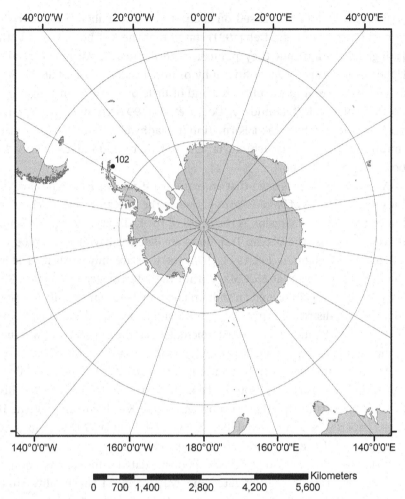

Figure 9.5 Map of Antarctica showing position of the Seymour Island locality (number 102 in Table 2.1 and Appendix) on the Antarctic Peninsula.

boundary. Early reports on the palynology of Cretaceous and Paleocene rocks of Antarctica emphasized the record of marine dinoflagellate cysts, which showed no evidence of a major extinction event at the presumed position of the K–T boundary.

The only K–T boundary section that has been studied on the continent of Antarctica is on Seymour Island on the Antarctic Peninsula (Figure 9.5). The K–T boundary was approximately located within the López de Bertodano Formation on Seymour Island (locality 102 in Table 2.1 and Appendix), based on the highest stratigraphic occurrences of marine foraminifera and ammonites (Askin 1988a, 1988b). Askin (1988b) located the boundary palynologically using dinoflagellate cysts. Her samples also contained pollen and spores of terrestrial plants that had been transported into the marine paleoenvironment. She stated that these terrestrial palynomorphs indicated little change in Antarctic land plants across the K-T boundary. Apparently Askin was basing her view on the record in Antarctica when she expressed some skepticism regarding the emerging palynological record from North America, saying that the reports were divergent and conflicting. Askin (1990) plotted the stratigraphic ranges of 41 angiosperm pollen species through the K-T boundary interval. No obvious extinction level is evident in Askin's data, but three taxa known to disappear near the boundary in Australia and New Zealand also disappear near the end of the Cretaceous on Seymour Island. Askin invoked climate change to account for palynofloral turnover across the boundary.

Elliot *et al.* (1994) finally verified the position of the K-T boundary in Antarctica based in part on marine palynology (dinoflagellate cysts) with the discovery of an iridium anomaly at that level. The anomaly is less than 2 ppb but is about 40 times the background level. They concluded that their locality on Seymour Island provided no compelling evidence for mass extinction among terrestrial plants at the boundary. No fern-spore spike such as that later found in New Zealand was reported in Antarctica.

9.6   New Zealand

The K-T boundary in New Zealand has been known in marine rocks since the publication of Alvarez *et al.* (1980), and about 20 localities are now known (Hollis 2003). The Woodside Creek locality on the South Island of New Zealand was one of the three localities in the world where anomalous concentrations of iridium were first detected at the paleontologically defined K-T boundary – the level of extinction of marine foraminifera. The boundary in nonmarine rocks in New Zealand took a great deal longer to locate, and was not definitively established until more than 20 years later.

An early paper by Raine (1984) presented a palynostratigraphic zonation of nonmarine Cretaceous and Paleogene rocks in New Zealand. The K-T boundary corresponded to the Hamurian-Teurian stage boundary, the New Zealand equivalent of the Maastrichtian-Paleocene boundary of North America and Europe. No extraordinary changes in the palynoflora were evident across this boundary, however. The ranges of two species of angiosperm pollen, *Tricolpites lilliei* and *Tricolpites secarius*, define a palynostratigraphic zone boundary at this stratigraphic position. Using those criteria, the K-T boundary was located in a general way in the Rewanui Member-Goldlight Member sequence exposed along the railway line between the Moody Creek coal mine and the town of Greymouth, on the west coast of South Island.

The boundary was definitively located by Vajda *et al.* (2001), less by an extinction level among angiosperm pollen than by the presence of a fern-spore spike.

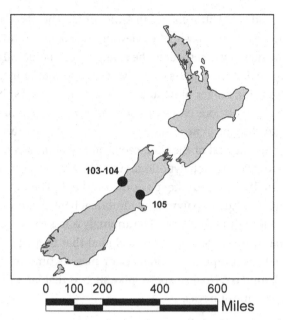

Figure 9.6 Map of New Zealand showing approximate positions of K–T boundary localities on the South Island (numbers 103–105 in Table 2.1 and Appendix).

Vajda *et al.* found a double fern-spore spike at two localities, a fully nonmarine site at the Moody Creek coal mine, and a second one in nearshore marine rocks along the Waipara River on the east coast of the South Island (Figure 9.6). The boundary is also marked by the extinction of two species of angiosperm pollen, *Tricolpites lilliei* and *Nothofagidites kaitangata*. A large iridium anomaly (71 ppb) is present at Moody Creek Mine (locality 103). At both Moody Creek Mine and Mid-Waipara River (locality 105), the fern-spore spike is composed of high percentages of *Laevigatosporites* and other spores, interpreted as having been produced by ground ferns, followed by *Cyathidites* and a second spore species, both interpreted as being from tree ferns (Figure 9.7). Tree ferns that produce spores of the morphology of *Cyathidites* are common in the modern flora of New Zealand; thus, the interpretation of the fossil spores above the K–T boundary as representing tree ferns is quite reasonable, although that inter-pretation has not been made for the ferns that produced the North American fern-spore spikes dominated by *Cyathidites* spores.

    In 2003, a special issue of the *New Zealand Journal of Geology and Geophysics* published several papers on the K–T boundary in both marine and nonmarine rocks in New Zealand. Six of the 20 K–T boundary sections now known in New Zealand were reviewed by Hollis (2003). Vajda and Raine (2003) and Hollis and Strong (2003) discussed the Mid-Waipara River locality in considerable detail. This locality is in nearshore marine rocks, and the microfossil record includes

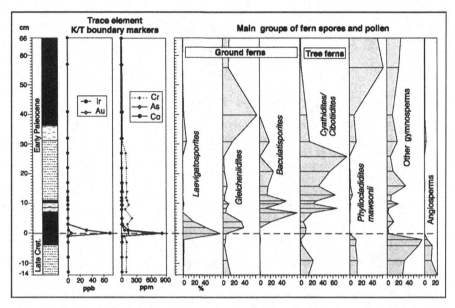

Figure 9.7 Fern-spore spikes above the K–T boundary in New Zealand (from Vajda *et al.* 2001). Copyright 2001 AAAS; reprinted by permission.

pollen and spores of terrestrial origin as well as marine dinoflagellate cysts, foraminifera, radiolarians, and calcareous nannofossils. In addition, the site preserves Cretaceous marine reptiles in large concretions not far below the K-T boundary horizon. The pollen and spore record at the Mid-Waipara River locality documents the existence of a temperate rainforest composed largely of podocarpaceous conifers and tree ferns in latest Maastrichtian time, which was destroyed by the K-T boundary event. Ground ferns temporarily replaced the forest vegetation. Both gymnosperms and angiosperms returned after an estimated several thousand years and reestablished the forest. Vajda and Raine (2003) presented detailed occurrence data and photomicrographs of the palynomorphs. They assert that the turnover in vegetation at the K-T boundary is comparable to that documented by palynological records from North America and Japan (see Sections 7.8 and 8.3). In their description of the Mid-Waipara River locality, Hollis and Strong (2003) noted that a weak iridium anomaly is present; but suggested that it is relatively weak because of post-depositional bioturbation of the sediments in this neritic paleoenvironment.

A third K–T boundary locality in New Zealand with nonmarine palynomorphs was reported by Vajda *et al.* (2004). Compressor Creek (locality 104) adds to the records of fern-spore spikes from the Southern Hemisphere. Vajda and McLoughlin (2007) summarized the features of the New Zealand palynofloral record at the K–T boundary and discussed using it as a tool for determining causes

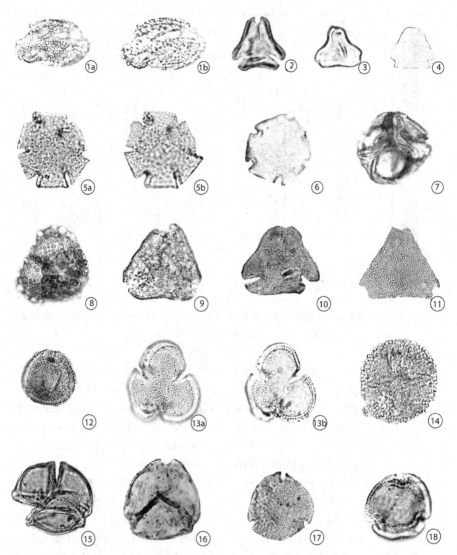

Figure 9.8 Some key pollen species from the Mid-Waipara locality in New Zealand (number 105 in Table 2.1 and Appendix), some specimens photographed at two levels of focus. 1a, 1b – *Ilexpollenites clifdenensis*; 2 – *Tricolpites secarius*; 3 – *Triorites minor*; 4 – *Triorites minor*; 5a, 5b – *Nothofagidites kaitangata*; 6 – *Nothofagidites waipawaensis*; 7 – *Dicotetradites clavatus*; 8 – *Paripollis* sp.; 9 – *Peninsulapollis askiniae*; 10 – *Peninsulapollis gillii*; 11 – *Proteacidites* sp.; 12 – *Proteacidites palisadus*; 13a, 13b – *Tricolpites reticulatus*; 14 – *Quadraplanus brossus*; 15 – *Tricolpites lilliei*; 16 – *Tricolpites lilliei*; 17 – *Tricolpites phillipsii*; 18 – *Tricolporites* sp. Specimens range from 20 to 55 micrometers in diameter. From Vajda and Raine (2003); reprinted by permission.

of the end-Permian mass extinction. Representatives of the uppermost Cretaceous and lowermost Paleocene palynofloras of New Zealand are shown in Figure 9.8.

Discovery of fern-spore spikes at the otherwise palynologically subtle K–T boundary in New Zealand has important implications for the history of plants at

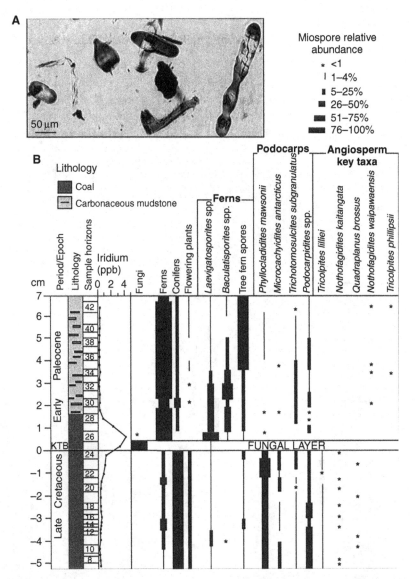

Figure 9.9 Fungal-spore spike above the K–T boundary in New Zealand (from Vajda and McLoughlin 2004). Copyright 2004 AAAS; reprinted by permission.

the K–T boundary. From the data presented by Vajda *et al.* (2001), it is apparent that effects of the K–T boundary event on plants were truly global in extent. Still more interesting data on the effects of the K–T boundary event on plants comes from the paper by Vajda and McLoughin (2004). They found abundant fungal spores in a 4-mm-thick layer at the Moody Creek Mine locality, coincident with the iridium anomaly and below the level of the fern-spore spike (Figure 9.9). They interpreted this fungal-spore spike, if it may be so-called, to represent

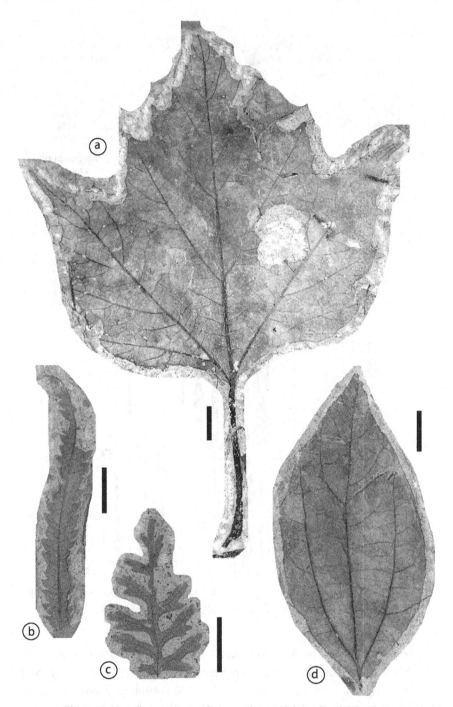

Figure 9.10 Paleocene leaves from northwest Nelson, South Island, New Zealand.
a – (DMNH 6654) b – Proteaceae (DMNH 6611) c – (DMNH 6692) d – Lauraceae
(DMNH 6640). Scale bar is 1 cm.

saprophytes existing on formerly photosynthetic vegetation that was destroyed by effects of the impact event.

It is quite interesting to note that in his early work on Upper Cretaceous and Paleogene palynostratigraphy, Raine (1984) was building on the earlier work of Couper (1960), who elucidated the stratigraphy and palynofloral succession in New Zealand. Couper correlated the terrestrial palynofloras of the Maastrichtian and Danian equivalents (the Hamurian and Teurian stages, respectively) with marine foraminiferal biostratigraphy, thereby laying the groundwork for locating the K–T boundary in the nonmarine strata of New Zealand. Some 40 years later, the research of Vajda *et al.* (2001) pinpointed the boundary in those rocks and made the major discoveries that contribute to our understanding of the global effects of the K–T boundary event on plants.

The megafloral record of the K–T boundary in New Zealand is being developed in two areas. One of us (KRJ) made collections near the Moody Creek Mine and Compressor Creek sites in 1992, retrieving fossil leaves from several locations in the Rewanui and Goldlight members. This area is heavily forested and outcrops were only accessible in road and river cuts, a situation that prevented precise correlation between localities. The Rewanui Member is primarily a fluvial sequence composed of sandstone, mudstone, and coal beds, whereas the Goldlight Member is a relatively uniform, dark gray mudstone of lacustrine origin. Because of this distinct difference in depositional environment, the small size of the collections, and the difficulty of correlating the localities, this work remains unpublished.

Kennedy *et al.* (2002) described Maastrichtian and Paleocene megafloras from the northwest tip of the South Island in Nelson. Here the Pakawau and Kapuni groups are exposed in roadcuts and have yielded well-preserved leaf impressions and palynomorphs. To date, this work has generated paleoclimate estimates for individual localities but has not been sampled with enough density to measure megaflora biostratigraphic trends. Some representatives of the Paleocene megaflora of New Zealand are shown in Figure 9.10.

PART III INTERPRETATIONS

10

Assessment of the K–T boundary event

In this book we have attempted to provide an overview of the state of knowledge of plants and the K–T boundary. The center of North America has yielded a rich floral record and the high quality and vast extent of exposures hold much promise for further refinement. Nearly all of the K–T boundary sections that contain evidence of the impact event are in North America. For this reason, it is difficult to make global generalizations other than to say that more sections are needed on other continents. In fact, it is fair to say that without the North American record, we would be hard pressed to argue for major floral change at the K–T boundary. Only the New Zealand sections document floral change relative to the iridium anomaly, and in those sections it is primarily the fern-spore spike that supports the concept of an impact as a causal mechanism. Were the New Zealand sections studied in isolation, it is questionable whether the fern-spore spike would have been recognized as significant. Only when taken in context of the North American fern-spore spike is the New Zealand occurrence interpretable. Nonetheless, its presence and close association with an iridium spike is one of the strongest arguments for the global reach of the immediate and deleterious effects of the bolide impact. The loss of forest canopy at the K–T boundary in New Zealand was followed by the nearly complete recovery of the pre-existing Cretaceous forest. This pattern, so different from the devastation seen in the North American localities, is also an argument for the attenuation of damage with increasing distance from the impact site.

It is clear that a highly resolved stratigraphy is a prerequisite for even beginning to study the K–T boundary in terrestrial rocks. It is a tall order to request stratigraphic precision on the order of the impactite level of resolution, and it can be achieved only after basic geologic mapping has defined the appropriate units and placed them in an accurate temporal framework. Palynology is the only method that allows for both regional assessment and microstratigraphic resolution, and it is clear that this field is where future breakthroughs in locating new K–T boundary sections will take place. But palynology alone is not enough. It must be coupled with independent radiometric or magnetostratigraphic measures of geologic time. The study of the K–T boundary has raised the bar of stratigraphic resolution to almost impossibly high standards. Where once stages or subchrons were considered high resolution, the presence of the K–T impactite demands resolution on the order of a few millimeters, or a few years. Essentially all non-North American terrestrial K–T boundary sections are wanting when it comes to this resolution. Thus, it is premature to draw conclusions from these records.

For this reason, our summary will focus primarily on the North American record and only refer to a few key points in other continents. For North America, palynology and megafloral paleobotany tell different parts of the same story. The palynology provides the regional spread of data and the high-precision correlation with the impact layer. Megafloral paleobotany is much more localized but does a better job of resolving patterns of climate, magnitude of extinction, plant–insect interactions, and plant–facies relationships.

The abruptness of the K–T palynological extinctions is unprecedented and surprising in the fossil record. There is no other place in the terrestrial geological record where dramatic and regionally extensive floral change occurs over a stratum measured at the millimeter level. This high level of precision and resolution is coupled with an equally high level of extinction that is also regional to continental in its extent. Depending on your perspective, the fact that the best terrestrial K–T record is located on the same continent as the Chicxulub impact site may seem like an unlikely, perhaps suspicious coincidence. To us, it seems a lucky coincidence that the Chicxulub impactor hit a continent that was undergoing orogeny and subsidence. The resulting prevalence of Late Cretaceous and Paleogene Laramide foreland basins made ready receptacles for airborne debris. The odds of the rock record preserving a specific centimeter of sediment seems vanishingly small, and the repeated occurrence of this low probability event is a testimony to the rapid subsidence of the Laramide basins.

The sedimentology of the K–T boundary still holds some unexplained patterns. Roland Brown's formula for identifying the K–T boundary, the first persistent lignite above the highest dinosaur bone, remains effective today and continues to raise valid questions about the relationship of sedimentary facies and the K–T boundary. Fortunately, the combination of palynology and the iridium anomaly have provided the resolution needed to decouple the initiation of mire formation and the evidence for an impact. Localities like Sugarite in the

Raton Basin and Pyramid Butte in North Dakota show clear evidence that the impact layer was deposited well after the formation of the "boundary lignite." In several sections north of Marmarth, North Dakota, the basal 2–3 m of the Fort Union Formation contains Cretaceous palynomorphs and at one locality, a *Triceratops* skull. Clearly the transition to mire-forming landscapes had come about in places before the Chicxulub impact occurred. In the central Great Plains, the initiation of mire formation seems linked to the transgression of the Cannonball Sea. This proximity to the K–T boundary is close enough that some authors have even gone as far as to suggest that marine regression caused the extinction of the dinosaurs (Archibald 1996). We feel that the marine regression hypothesis as an explanation for the dinosaur extinction was based on a very coarse reading of the stratigraphic record of the boundary. In other areas like Alberta and Colorado, where there is no evidence of an early Paleocene seaway, a lignite bed is often but not always associated with the impact bed, suggesting that there are other reasons for the initiation of mires at this time.

The occurrence of the Fort Union zero (FU0) megaflora in the Fort Union strata of Cretaceous age in North Dakota provides possible insight about the ultimate floras of the Cretaceous. This megaflora is associated with a Cretaceous palynoflora but is represented by a depauperate leaf assemblage containing many of the species that become dominant in the Paleocene. When first discovered, it was quite confusing. It is now interpreted as a Cretaceous mire flora. The discovery of this flora suggests that the Paleocene landscape was disproportionately populated by plants that had lived in Cretaceous swamps and that drier-ground Cretaceous taxa were the forms that suffered more extinction. This observation is similar to what is seen in the vertebrate communities at the K–T boundary where aquatic forms including turtles and crocodiles fared far better than did terrestrial forms such as dinosaurs and mammals. This hypothesis needs further testing because no FU0 megafloral assemblages have been discovered more than three meters below the K–T boundary.

Another fascinating and subtle sedimentological feature of the K–T boundary transition is the apparent decrease in pedogenic alteration of fine-grained sediments above the boundary. This is clear in the central Great Plains where the well-rooted paleosols of the Hell Creek Formation give way to the variegated and laminated siltstone deposits of the Fort Union Formation. In the Denver Basin, a similar transition occurs within the Denver Formation at the K–T boundary. The result in both cases is that fossil leaves are more commonly preserved in the Paleocene portion of the section than they are in the Cretaceous portion. This is in part responsible for the historical over-collection of Paleocene fossils relative to Upper Cretaceous fossils.

Casual observation of this pattern suggests that early Paleocene landscapes were relatively denuded in comparison to Late Cretaceous ones, but this explanation seems too facile and perhaps represents too literal a reading of the rock record. Certainly the high-resolution records afforded by palynology suggest that the duration of post-impact herbaceous vegetation represented by the fern-spore spike was short lived, lasting only as long as it took to deposit the few centimeters of strata that represent the spike. The transition from heavily rooted Cretaceous strata to well-bedded Paleocene strata seems to have been much more long-lasting. Because stratigraphy is local, the fact that this pattern is present on a continental scale suggests that something more than local depositional conditions may have been responsible. In addition to the fern-spore spike, the coal-petrographic record from the Wood Mountain core in Saskatchewan argues strongly for the devastation of the forest canopy immediately after the K–T boundary and its relatively rapid recovery within the time needed to accumulate ~50 cm of lignite.

What begins to become clear is a pattern of rapid deforestation at the K–T boundary followed by rapid reforestation by a depauperate group of swamp-loving trees. The distribution of fossil sites along an extensive latitudinal transect from New Mexico to Alberta provides the beginnings of a database where it will be possible to look at the latitudinal ranges of taxa before and after (in the case of survivors) the impact. Some initial observations on this topic are offered.

One first-order observation is that plants producing *Aquilapollenites* were more speciose in the northern Plains and Canada than they were in the Raton Basin to the south. Because this group suffered nearly complete extinction at the K–T boundary, the percent of extinction decreased from north to south. This counters the expectation that extinction would be more extreme the closer the sample locality is to the Chicxulub crater. Perhaps this expectation is too simple. If the immediate effects of the asteroid impact included a blow-down of the standing forest followed by fire and then months of darkness and acid rain, we might expect that survivorship had more to do with ability to regenerate from rootstocks or seed banks rather than any advantage accrued by simple distance from the blast.

We are just now beginning to have enough geographically distributed data to look at the differential response of specific taxa with latitude, and several patterns are beginning to emerge. One example is the lineage of plants that have improperly been attributed to the genus *Artocarpus*. "*Artocarpus*" *lessigiana* and related species were present in the last three million years of the Cretaceous and the first two million years of the Paleocene in the Denver and Raton basins, but only the last 500 000 years of the Cretaceous in the Williston Basin. *Platanites marginatum* and "*Zyziphus*" *fibrillossus* also share this pattern of occurrence. One possible explanation for this pattern is that certain species were limited by temperature and were able to extend their range to the north only under certain conditions. The cycad *Nilssonia* existed only in the last 500 000 years of the Cretaceous in the Williston Basin and is not present at all in the Denver and Raton basins. These patterns suggest that floral devastation was localized and that recovery was mediated by local climate. Plants that were marginal in a landscape might have been vulnerable to local extirpation by the impact event, and were not able to reestablish themselves in those areas because of the properties that made them marginal in the first place.

It is not clear how many kinds of plant refugia existed, but certainly there must have been a variety. We have already suggested that mire plants seem to have survived better than dry-land plants in the Great Plains. It is also plausible that proximity to mountain ranges may have been beneficial due to the variety of habitats and potentially blast-sheltering topographies. In the case of the 1980 Mount St. Helens eruption and subsequent nuée ardente, topography was a significant factor in determining which trees were flattened and which remained standing. We have begun to pursue these questions in the Denver Basin by mapping Cretaceous and Paleocene floral occurrences relative to the margin of the Colorado Front Range to define vegetational trends relative to topography. Recent discoveries of Paleocene megafloras in the Middle Park Formation, which is located in the axial Middle Park Basin on the west side of the Front Range, have shown evidence for relative topography in the Rockies shortly after the K–T boundary. Prospecting in Cretaceous rocks in this area has the potential to define Cretaceous montane floras and perhaps to define K–T boundary refugia.

11

Evaluation of scenarios for the K–T boundary event

11.1 Overview

Several scenarios for events at and following the K–T boundary have been proposed, some of them predating the one set forth in Alvarez *et al.* (1980), the publication that substantially renewed speculation on the subject. In this chapter we summarize the most prominent hypotheses, including that advanced by the Alvarez team. In the final section of this chapter, we discuss our own preferred scenario, which is based on our understanding of effects of the boundary event on plant life.

Other reviews of K–T boundary scenarios, most of which emphasize the impact hypothesis, are presented in books by Albritton (1989), who also reviewed the whole concept of catastrophic events in Earth history and the way they have been viewed since the seventeenth century; Allaby and Lovelock (1983); Alvarez (1997); Frankel (1999); Hsü (1986); and Powell (1998). Archibald (1996) and Officer and Page (1996) proposed alternative scenarios to account for the mass extinctions at the end of the Cretaceous. In general, these books feature the history of dinosaurs and tend to neglect the fossil plant record, exceptions being Hsü (1986) and Powell (1998); some of them overlook plants entirely. This short shrift or omission from prominent books on the subject served as our inspiration for writing this book.

Disregarding a few unconventional and even unscientific (in the sense of being untestable) ideas about causes of the K–T extinctions (which also focus on dinosaurs), the mainstream hypotheses center on four themes: climate change, regression of epicontinental seas, a major episode of volcanic eruptions, and impact of an extraterrestrial body. We briefly discuss and critique each of these in the following sections.

11.2   Climate change

Climate change has been a popular explanation for extinctions from the time that the phenomena of mass extinctions in the geologic record and major climatic fluctuations in the geologic past (such as those during the ice ages of the Pleistocene) were first recognized. Thus, it is an old idea, but the one still favored by our Russian colleagues to account for the extinction of plants at the K–T boundary (see Section 8.5). Climate change would seem to be the default answer for vegetation change, but there is a burden of proof that is lacking. It is not adequate simply to assume that climatic deterioration from the hothouse world of the Cretaceous toward the icehouse world of the Pleistocene can explain the disappearance of plants at the K–T boundary in North America. The abrupt nature of those disappearances as documented in Chapters 6 and 7 make such a facile explanation untenable. As exemplified by changes in vegetation associated with the advance and retreat of continental glaciers during the Pleistocene, floras migrate rather than become extinct.

Fossil plants (leaves) contain a climate signal that has been interpreted for the K–T transition (e.g., Wolfe and Upchurch 1987b, Wilf and Johnson 2004); shifts in temperature that affected vegetation provide a background against which the sudden extinctions at the K–T boundary stand in sharp contrast. The essence of the issue is that climate change is gradual, but the extinctions at the K–T boundary, at least in North America, were abrupt. Furthermore, on the global scale there is the correlation between land and sea that ties the terrestrial K–T boundary to the marine record and effectively separates climate change from the impact event (Huber *et al.* 1995, Wilf *et al.* 2003).

11.3   Regression

Retreat of the Western Interior seaway from the continent at the end of the Cretaceous is the explanation for extinctions on land favored by Archibald (1996), and as noted by Powell (1998), it is the most-cited cause of dinosaur extinction. Archibald's idea was that withdrawal of the sea (regression) would somehow break up habitats on land and that "habitat fragmentation" would hinder the flow of populations of animals and plants between isolated parts of their former habitat, leading to their extinction. We critique Archibald's idea below, but first, some general comments about the relationship (or lack thereof) between regression and extinction.

Haq *et al.* (1988) published a now-famous curve depicting major sea-level fluctuations (transgression and regression) during the past 542 million years of Phanerozoic time. The simple fact is that none of the major episodes of extinction during that time – including that at the K–T boundary – correlate with major episodes of regression. The sea-level curve of Haq *et al.* also indicates that there were many substantial fluctuations in sea level during the Cretaceous, none of which are associated with extinctions in either the marine or terrestrial realms. In more recent geological time, the major changes in sea level during the Pleistocene were not accompanied by major extinctions of land plants or animals. Furthermore, much like the process of climate change, regressions take place over thousands of years; they are not abrupt, as are the extinctions of animals and plants at the K–T boundary. In any particular area, one would expect the responses of organisms to major changes in sea level to be time-transgressive, yet the record of plants in North America documented in Chapters 6 and 7 is one of continent-wide, geologically instantaneous devastation. This fact alone should render the regression–extinction scenario as unsubstantiated and hard to defend.

With particular reference to North America, in Archibald's study area we find that his hypothesis does not stand up against the detailed record of changes in sea level during the Maastrichtian and the Paleocene. The major regression during the latest Cretaceous in this region was essentially complete 3–4 million years before the K–T boundary event (Johnson *et al.* 2002), but during the rest of Maastrichtian time, rich Cretaceous floras and faunas (including dinosaurs) dominated the landscape. Furthermore, at the very time that "habitat fragmentation" due to marine regression was allegedly taking place, the Western Interior seaway was readvancing into the North Dakota area in minor transgressive pulses represented by the Breien Member and the Cantapeta Tongue of the Hell Creek Formation (Hoganson and Murphy 2002). That last transgression of marine waters into the interior of the North American continent culminated with the Cannonball sea, which is represented by the Cannonball Member of the Fort Union Formation (see Figure 6.2). Thus, the habitat-fragmentation hypothesis of Archibald (1996) has no evidential support in the Cretaceous of western North America, source of the world's richest dinosaur faunas and most intensely studied megafloras and palynofloras.

11.4   Volcanism

The scenario favored by Officer and Page (1996) involves massive volcanic eruptions that might have spewed sunlight-obscuring dust into the atmosphere initiating a global crisis like that caused by the extraterrestrial impact envisioned by Alvarez *et al.* (1980) (see Section 11.5). Poisonous gases such as chlorine and sulfur dioxide are also part of volcanic eruptions. The Officer and

Page volcanic hypothesis is essentially an alternative to the Alvarez *et al.* impact hypothesis. Its major flaw is the timing of the eruptions.

The location of the greatest of the volcanic eruptions around K–T time discussed by Officer and Page is in India; the lavas that flowed formed the Deccan Traps (see Section 9.4). As discussed in Section 9.4, the basaltic lava of the Deccan Traps erupted during an interval of a few million years, beginning long before the end of Cretaceous time and continuing into earliest Paleocene time. In contrast, the K–T extinctions were abrupt; by all possible measurement, they were geologically instantaneous. As with the hypothesis of climate change failing because it is inconsistent with the record of abrupt extinction at the K–T boundary, no meaningful linkage can be found between volcanism and the K–T extinctions. Furthermore, the possibility that volcanic emanations darkened the skies and thereby lowered atmospheric temperatures is falsified by the data of Huber *et al.* (1995) and Wilf *et al.* (2003), who show a warming trend associated with Deccan volcanism. In short, although major volcanic eruptions can have devastating effects locally and be the cause of local extirpations of flora and fauna, there is no evidence that they do so on a global scale or specifically that they did so at the K–T boundary.

11.5   Impact scenario

The extraterrestrial impact scenario for the extinctions at the K–T boundary introduced by Alvarez *et al.* (1980) is the one that is best supported by more than 25 years of research. The evidence for the impact having occurred is overwhelming. It includes the globally distributed, anomalous concentration of iridium at the K–T boundary (Alvarez *et al.* 1980, Alvarez *et al.* 1984); shocked quartz and other mineral grains produced at the impact site and deposited at the boundary (Bohor *et al.* 1984, Bohor *et al.* 1987b, Izett 1990); and microtektite spherules generated by the impact and deposited in the boundary claystone layer (Smit 1999). Furthermore, the impact crater has been found (Hildebrand *et al.* 1991). Physical effects of the K–T impact event on the Earth are well described by Covey *et al.* (1994), Toon *et al.* (1994), Pope *et al.* (1997), and Smit (1999) and are briefly summarized below. The data presented in Chapters 6 and 7 of this book address what at the outset we termed the "Alvarez challenge," to determine whether the fossil record of plants supports the hypothesis that the impact caused the K–T extinction. We conclude that in western North America it does.

The essential aspect of the Alvarez *et al.* (1980) impact hypothesis as it relates to the fate of vegetation is darkness. That is, after the impact, solar radiation was blocked to the extent that photosynthesis was shut down. This effect would kill plants within a short time, perhaps only weeks. Some plant species would survive because they could regenerate from root stocks or seeds already in the ground, but others would not. Entire ecosystems would collapse with the destruction of plant communities. According to the impact scenario, darkening of the skies was caused by dust and sulfate aerosols injected into the stratosphere by the impact; global wildfires (discussed below) might also have contributed smoke and soot, and soot would have absorbed sunlight more effectively than rock dust. The cold of an "impact winter" would ensue. Our review of the global reaction of vegetation at the K–T boundary shows that plant communities in North America were affected in a way consistent with this mechanism, but outside North America, the evidence is less clear or lacking. This observation may be in agreement with the theoretical calculations of Pope (2002), who concluded that clastic debris from the impact would tend to spread across North America, the Pacific, and Europe, but that little would reach the Southern Hemisphere. The records of plant extinction at the K–T boundary in Europe are ambiguous (see Section 8.2), but the record from New Zealand (Vajda et al. 2001) strongly suggests that there was at least a transitory effect at that latitude (see Section 9.6). In that part of the world, sulfate aerosols and soot from wildfires may have caused shutdown of photosynthesis.

The possibility that global wildfires were set by the impact due to radiation from the fireball and the expanding cloud of rock vapor was proposed by Wolbach et al. (1985, 1990). Ignition of wildfires by ballistic re-entry of ejecta from the impact was proposed by Melosh et al. (1990) and vigorously supported by Robertson et al. (2004). However, direct stratigraphic evidence for these wildfires is equivocal. Wolbach et al. (1985) described graphitic carbon in samples of the K–T boundary claystone from Denmark, Spain, and New Zealand, evidence that was accepted by Melosh et al. (1990). On the other hand, as discussed in Section 7.6, McIver (1999) saw no evidence of wildfires at or above the K–T boundary in Saskatchewan, and Belcher et al. (2003) found no evidence of anomalous charcoal at the boundary at localities they studied in Colorado, Montana, North Dakota, Wyoming, and Saskatchewan. We have not observed unusual concentrations of charcoal, which is derived from charred plant material, in our extensive studies of palynological preparations from numerous K–T boundary localities in North America. While the impact model seems to predict fires, we are impressed by the apparent lack of direct evidence for them in the well-constrained sections of North America. This having been said, the presence of the fern-spore spike suggests that forests were knocked over on a continental scale, raising the possibility that there may have been a major blast effect independent of any major burn.

The possibility that the K–T boundary impact event resulted in unprecedented acid rain composed of nitric and/or sulfuric acids has been proposed and discussed widely in the literature. Key references include Prinn and Fegley (1987), Brett (1992), Sigurdsson *et al.* (1992), D'Hondt *et al.* (1994), and Pope *et al.* (1997). There is little or no direct evidence on this subject available from the record of fossil plants.

There are other, less-often discussed side effects of the K–T boundary impact that significantly influenced plants. There is good evidence that the extinctions included insects (Labandeira *et al.* 2002). The loss of insect herbivores must have profoundly affected plant communities of the earliest Paleocene in North America. The possible loss of pollinating insects (Sweet and Jerzykiewicz 1987, Frederiksen *et al.* 1988, Sweet and Braman 2001) remains speculative, although it is true that, in the early Paleocene, North American floras were dominated by wind-pollinated plants, such as angiosperms that produced Normapolles pollen or, at northern latitudes, taxodiaceous/cupressaceous gymnosperms.

It may well be that some plant species survived the extinction event by occupying refugia, isolated locations where relict populations were able to avoid the destructive consequences of the impact. With reference to the K–T boundary, the concept of refugia was introduced by Tschudy and Tschudy (1986) to account for survival of some species that were severely reduced in abundance at the boundary but persisted into the Paleocene. The exact nature of such refugia is unknown, but topography and protection from blast effects of the impact may have been critical. As mentioned in Section 6.2, mires may have provided respite for some species because, in western North America, survivorship appeared to be greatest among plants that had inhabited mires (Johnson 2002). Perhaps refugia played a role in the early Paleocene development of rainforests in Colorado (Johnson and Ellis 2002; Ellis *et al.* 2003), where diverse species thrived in a location in close proximity to the Rocky Mountain Front at the same time that floras in distal locations remained depauperate well into Paleocene time.

11.6   The preferred scenario

Our preferred scenario for the K–T boundary event focuses on the extraterrestrial impact hypothesized by Alvarez *et al.* (1980) and authenticated by investigators in many fields in later years (references in Section 11.5, above). With regard to the record of fossil plants, western North America, the region with the highest resolution, shows a clear pattern of major extinction followed by gradual recolonization by a significantly altered flora. Other regions of the world await higher resolution stratigraphy and/or better sampling for this pattern to become as clear, or for a different pattern to emerge.

In western North America, the Maastrichtian was characterized by a varied and interesting angiosperm-dominated landscape. At the moment of impact there was immediate knock-down of forests (McIver 1999). Possibly the fallen vegetation burned, but the data are equivocal. The earliest Paleocene saw rapid colonization of the landscape by spore-bearing plants (ferns and sphagnum moss) as described by Tschudy *et al.* (1984), and possibly at northern latitudes by certain angiosperm species (Sweet and Braman 1992). This was followed by continent-wide colonization by swamp-loving plants, the FUI flora of Johnson and Hickey (1990). However, some plants that existed before the extinction did survive. What transpired was that an existing biota was struck down and the survivors became the colonizers of the Paleocene. Refugia may have promoted early development of rainforests such as that at Castle Rock, Colorado (Johnson and Ellis 2002; Ellis *et al.* 2003). The plant communities that flourished in the Paleocene evidently were largely free of insect herbivores (Labandeira *et al.* 2002, Wilf *et al.* 2006). Vertebrate herbivores that preyed on plants were an entirely different stock from that which had existed during the latest Maastrichtian because the extinction event involved the loss of all large-bodied terrestrial vertebrates (Pearson *et al.* 2001, 2002).

Floral effects of the K–T boundary event

At the scale of western North America, the paleobotanical record appears to document a mass extinction of plant species over the time represented by a centimeter of sediment. Before 1980, temporal resolution of the rock record was coarse and no such statement could have been made with any sense of certainty. The K–T boundary provided a testable hypothesis that is still being examined. Our results seem to suggest that Traverse (1988a) was incorrect in asserting that all changes in floras throughout time have been due to gradual replacement. It now appears that it is possible for plants, like animals, to suffer abrupt mass extinction.

The radiation of the angiosperms that began at the start of the Cretaceous was well under way by the end of the period some 80 million years later and most terrestrial ecosystems either contained or were completely dominated by angiosperms. The K–T extinction, while selective at the environmental scale as evidenced by the apparent preferential survival of mire plants, was not obviously selective at a taxonomic level. That is to say, no major plant groups disappeared at the boundary, and the damage primarily occurred at the species level as local ecosystems independently suffered the after-effects of the impact. Floral recovery from the event seems to have taken the duration of the Paleocene with the awkward exception of the Castle Rock rainforest, whose presence less than two million years after the event remains enigmatic.

This leads us to the interesting question of whether the K–T event had a destructive or formative effect on entire biomes. Although the relationship of large dinosaurian herbivores to their associated vegetation will always be difficult to qualify, it is hard to imagine that they were not keystone species in some sense of the phrase. The disappearance of non-avian dinosaurs must have been significant for their environment. Moving forward, it will be interesting to search for more evidence of Paleocene rainforests to see if their occurrence is globally tied to the K–T event or if the Colorado occurrence is an isolated one.

Clearly, the fossil record of plants at the K–T boundary, while strong in North America, is weak on other continents. The sections in eastern Russia and northern China have fossils but lack continuous outcrops and independent dating. The sections in Patagonia are promising but largely unexplored, and the floras of New Zealand await further collecting and analysis. All these areas need more stratigraphic framework and larger plant collections before they can contribute substantially to this story. The new discoveries in Colombia are perhaps the most promising because they occur directly on the paleoequator and promise significant information about the tropics.

The potential for these areas is high if future paleontologists take advantage of the geologically unique K–T boundary impactite. This thin layer is perhaps the most precise and global time marker in the entire geologic record, and it is as much a tool for future research as it is the record of global devastation in the past.

Appendix

Appendix*

Locality name	(1) Pyramid Butte	(2) Sunset Butte	(3) River Section	(4) Bobcat Butte	(5) Mud Buttes
Latitude	46° 25′ 03″ N	46° 08′ 59″ N	46° 20′ 57″ N	46° 17′ 42″ N	46° 01′ 33″ N
Longitude	103° 58′ 33″ W	103° 48′ 37″ W	103° 53′ 48″ W	103° 52′ 46″ W	103° 52′ 30″ W
County	Slope	Bowman	Slope	Slope	Bowman
State/Province	North Dakota	North Dakota	North Dakota	North Dakota	North Dakota
Country	USA	USA	USA	USA	USA
Formation(s)	Hell Creek and Fort Union	Hell Creek and Fort Union	Hell Creek and Fort Union	Hell Creek and Fort Union	Hell Creek and Fort Union
Basin	Williston	Williston	Williston	Williston	Williston
Principal reference	Johnson et al. (1989)	Nichols and Johnson (2002)	Nichols and Johnson (2002)	Nichols and Johnson (2002)	Nichols and Johnson (2002)
Sub decimeter palynology resolution	Yes	No	No	Yes	Yes
Sampled interval (m)	1.9	111.5	30.3	76.2	12.6
K samples for palynology	6	18	7	17	6
T samples for palynology	6	8	3	1	5
Fern-spore spike	No	No	No	No	Yes
Iridium anomaly	0.72 ppb	No	No	No	1.38 ppb
Claystone layer	No	No	No	No	Yes
Shocked minerals	Yes	No	No	No	Yes
Magnetostratigraphy	Yes (1)	Yes (1)	Yes (1)	Yes (1)	Yes (1)
Radiometric dates	No	No	No	No	No
K vertebrates below	Yes (2, 3)	Yes (2, 3)	Yes (2, 3)	Yes (2, 3)	Yes (2, 3)
T vertebrates above	Yes (4)	Yes (4)	Yes (4)	Yes (4)	Yes (4)
K leaves below	Yes (5, 6)	Yes (5, 6)	Yes (5, 6)	Yes (5, 6)	Yes (5, 6)
T leaves above	Yes (5, 6)	Yes (5, 6)	Yes (5, 6)	Yes (5, 6)	Yes (5, 6)

* NB: Data are from principal reference unless otherwise cited. Other relevant references, numbered in parentheses, are listed at the end of the Appendix.

Locality name	(6) Pretty Butte	(7) Cannonball Creek	(8) Big Boundary	(9) Pretty Butte North	(10) Terry's Fort Union Dinosaur
Latitude	46° 22′ 43″ N	46° 27′ 37″ N	46° 23′ 30″ N	46° 23′ 04″ N	46° 28′ 01″ N
Longitude	103° 56′ 51″ W	104° 00′ 08″ W	103° 57′ 40″ W	103° 57′ 06″ W	104° 00′ 28″ W
County	Slope	Slope	Slope	Slope	Slope
State/Province	North Dakota	North Dakota	North Dakota	North Dakota	North Dakota
Country	USA	USA	USA	USA	USA
Formation(s)	Hell Creek and Fort Union	Hell Creek and Fort Union	Hell Creek and Fort Union	Hell Creek and Fort Union	Hell Creek and Fort Union
Basin	Williston	Williston	Williston	Williston	Williston
Principal reference	Nichols and Johnson (2002)	Nichols and Johnson (2002)	Nichols and Johnson (2002)	Nichols and Johnson (2002)	Nichols and Johnson (2002)
Sub decimeter palynology resolution	No	No	No	No	Yes
Sampled interval (m)	60.6	2.1	15.0	7.0	2.7
K samples for palynology	6	7	17	10	16
T samples for palynology	11	2	5	1	1
Fern-spore spike	No	No	No	No	No
Iridium anomaly	No	No	No	No	No
Claystone layer	No	No	No	No	No
Shocked minerals	No	No	No	No	No
Magnetostratigraphy	Yes (1)	Yes (1)	Yes (1)	Yes (1)	Yes (1)
Radiometric dates	No	No	No	No	No
K vertebrates below	Yes (2, 3)	Yes (2, 3)	Yes (2, 3)	Yes (2, 3)	Yes (2, 3)
T vertebrates above	Yes (4)	Yes (4)	Yes (4)	Yes (4)	Yes (4)
K leaves below	Yes (5, 6)	Yes (5, 6)	Yes (5, 6)	Yes (5, 6)	Yes (5, 6)
T leaves above	Yes (5, 6)	Yes (5, 6)	Yes (5, 6)	Yes (5, 6)	Yes (5, 6)

Locality name	(11) Mikey's Delite	(12) Dean's High Dinosaur	(13) Scorpion Stung Kristian	(14) Vertical Doug	(15) Badland Draw
Latitude	46° 28' 10'' N	46° 27' 39'' N	46° 27' 00'' N	46° 27' 31'' N	46° 26' 24'' N
Longitude	103° 59' 43'' W	104° 00' 21'' W	104° 01' 27'' W	103° 59' 22'' W	104° 00' 10'' W
County	Slope	Slope	Slope	Slope	Slope
State/Province	North Dakota	North Dakota	North Dakota	North Dakota	North Dakota
Country	USA	USA	USA	USA	USA
Formation(s)	Hell Creek and Fort Union	Hell Creek and Fort Union	Hell Creek and Fort Union	Hell Creek and Fort Union	Hell Creek and Fort Union
Basin	Williston	Williston	Williston	Williston	Williston
Principal reference	Nichols and Johnson (2002)	Nichols and Johnson (2002)	Nichols and Johnson (2002)	Nichols and Johnson (2002)	Nichols and Johnson (2002)
Sub decimeter palynology resolution	Yes	Yes	Yes	No	Yes
Sampled interval (m)	2.7	2.8	3.5	2.7	2.7
K samples for palynology	5	8	6	8	6
T samples for palynology	2	5	2	3	5
Fern-spore spike	No	No	No	No	No
Iridium anomaly	No	No	No	No	No
Claystone layer	No	No	No	No	No
Shocked minerals	No	No	No	No	No
Magnetostratigraphy	Yes (1)	Yes (1)	Yes (1)	Yes (1)	Yes (1)
Radiometric dates	No	No	No	No	No
K vertebrates below	Yes (2, 3)	Yes (2, 3)	Yes (2, 3)	Yes (2, 3)	Yes (2, 3)
T vertebrates above	Yes (4)	Yes (4)	Yes (4)	Yes (4)	Yes (4)
K leaves below	Yes (5, 6)	Yes (5, 6)	Yes (5, 6)	Yes (5, 6)	Yes (5, 6)
T leaves above	Yes (5, 6)	Yes (5, 6)	Yes (5, 6)	Yes (5, 6)	Yes (5, 6)

	(16) New Facet Boundary	(17) Torosaurus Section	(18) Huff	(19) Katus Site	(20) Brenner Site
Locality name					
Latitude	46° 27′ 34″ N	46° 05′ 33″ N	46° 36′ 37″ N	46° 00′ 09.9″ N	46° 03′ 58.5″ N
Longitude	103° 58′ 58″ W	103° 45′ 42″ W	100° 38′ 20″ W	101° 33′ 31.1″ W	101° 28′ 59.9″ W
County	Slope	Slope	Morton	Sioux	Grant
State/Province	North Dakota	North Dakota	North Dakota	North Dakota	North Dakota
Country	USA	USA	USA	USA	USA
Formation(s)	Hell Creek and Fort Union	Hell Creek and Fort Union	Hell Creek and Fort Union	Hell Creek and Fort Union	Hell Creek and Fort Union
Basin	Williston	Williston	Williston	Williston	Williston
Principal reference	Nichols and Johnson (2002)	Nichols and Johnson (2002)	Lerbekmo and Coulter (1984)	Murphy et al. (1995)	Murphy et al. (1995)
Sub decimeter palynology resolution	Yes	Yes	No	No	No
Sampled interval (m)	3.2	4.8	3.4	28.9	4.8
K samples for palynology	9	6	6	4	3
T samples for palynology	5	2	2	4	3
Fern-spore spike	No	No	No	No	No
Iridium anomaly	No	No	No	No	No
Claystone layer	No	No	No	No	No
Shocked minerals	No	No	No	No	No
Magnetostratigraphy	Yes (1)	Yes (1)	Yes	No	No
Radiometric dates	No	No	No	No	No
K vertebrates below	Yes (2, 3)	Yes (2, 3)	No	Yes	No
T vertebrates above	Yes (4)	Yes (4)	No	No	No
K leaves below	Yes (5, 6)	Yes (5, 6)	No	No	No
T leaves above	Yes (5, 6)	Yes (5, 6)	No	No	No

Locality name	(21) Cannonball Stage Stop Site	(22) Rattlesnake Butte Site	(23) Schaeffer Site	(24) Knispel Site	(25) Miller Site
Latitude	46° 13' 51.3" N	46° 09' 04.8" N	46° 01' 48.2" N	46° 03' 38.7" N	46° 36' 59.6" N
Longitude	101° 28' 3.9" W	101° 25' 52.4" W	101° 20' 06.4" W	101° 16' 49.4" W	100° 38' 19.6" W
County	Grant	Grant	Sioux	Sioux	Morton
State/Province	North Dakota	North Dakota	North Dakota	North Dakota	North Dakota
Country	USA	USA	USA	USA	USA
Formation(s)	Hell Creek and Fort Union	Hell Creek and Fort Union	Hell Creek and Fort Union	Hell Creek and Fort Union	Hell Creek and Fort Union
Basin	Williston	Williston	Williston	Williston	Williston
Principal reference	Murphy et al. (1995)	Murphy et al. (1995)	Murphy et al. (1995)	Murphy et al. (1995)	Murphy et al. (1995)
Sub decimeter palynology resolution	No	No	No	No	No
Sampled interval (m)	8.2	13	25	2.7	1.0
K samples for palynology	3	3	4	4	11
T samples for palynology	3	1	3	4	14
Fern-spore spike	No	No	No	No	No
Iridium anomaly	No	No	No	No	No
Claystone layer	No	No	No	No	No
Shocked minerals	No	No	No	No	No
Magnetostratigraphy	No	No	No	No	Yes (7)
Radiometric dates	No	No	No	No	No
K vertebrates below	No	No	No	No	No
T vertebrates above	No	No	No	No	No
K leaves below	No	No	No	No	No
T leaves above	No	No	No	No	No

Locality name	(26) Stumpf Site	(27) University of Mary Site	(28) Snyder Site	(29) Glendive	(30) Brockton
Latitude	46° 34′ 08.3″ N	46° 42′ 50.8″ N	46° 34′ 22.9″ N	47° 05′ 44″ N	48° 07′ 37″ N
Longitude	100° 37′ 15.6″ W	100° 45′ 53.9″ W	100° 29′ 1.87″ W	104° 41′ 52″ W	105° 00′ 13″ W
County	Morton	Burleigh	Emmons	Dawson	Richland
State/Province	North Dakota	North Dakota	North Dakota	Montana	Montana
Country	USA	USA	USA	USA	USA
Formation(s)	Hell Creek and Fort Union	Hell Creek and Fort Union	Hell Creek and Fort Union	Hell Creek and Fort Union	Hell Creek and Fort Union
Basin	Williston	Williston	Williston	Williston	Williston
Principal reference	Murphy et al. (1995)	Murphy et al. (1995)	Murphy et al. (1995)	Tschudy (1970)	Tschudy (1970)
Sub decimeter palynology resolution	No	No	No	No	No
Sampled interval (m)	3.6	3.2	3.9	113	52
K samples for palynology	2	2	3	18	3
T samples for palynology	3	3	1	2	5
Fern-spore spike	No	No	No	Yes (8)	No
Iridium anomaly	No	No	No	No	No
Claystone layer	No	No	No	No	No
Shocked minerals	No	No	No	No	No
Magnetostratigraphy	No	No	No	Yes (8)	No
Radiometric dates	No	No	No	No	No
K vertebrates below	Yes	No	No	Yes	Yes
T vertebrates above	No	No	No	Yes (8)	No
K leaves below	No	No	No	No	No
T leaves above	No	No	No	No	No

Locality name	(31) Brownie Butte [1970]	(32) Brownie Butte [1984]	(33) Brownie Butte [1999]	(34) Herpijunk Promontory	(35) Herpijunk Northeast
Latitude	47° 31′ 51″ N	47° 31′ 49″ N	47° 31′ 47″ N	47° 31′ 37″ N	47° 31′ 28″ N
Longitude	107° 00′ 59″ W	107° 01′ 02″ W	107° 00′ 21″ W	107° 12′ 13″ W	107° 11′ 47″ W
County	Garfield	Garfield	Garfield	Garfield	Garfield
State/Province	Montana	Montana	Montana	Montana	Montana
Country	USA	USA	USA	USA	USA
Formation(s)	Hell Creek and Fort Union	Hell Creek and Fort Union	Hell Creek and Fort Union	Hell Creek and Fort Union	Hell Creek and Fort Union
Basin	Williston	Williston	Williston	Williston	Williston
Principal reference	Tschudy (1970)	Tschudy et al. (1984)	Sweet et al. (1999)	Smit and Van der Kaars (1984)	Hotton (2002)
Sub decimeter palynology resolution	No	Yes	Yes	Yes	Yes
Sampled interval (m)	35	0.08	0.2	6.5	35.4
K samples for palynology	1	3	6	9	4
T samples for palynology	8	1	4	10	17
Fern-spore spike	No	Yes	Yes (13)	Yes	Yes
Iridium anomaly	No	1.04 ppb	Yes (13)	11.4 ppb	Yes (amount unspecified)
Claystone layer	No	Yes	Yes	Yes	Yes
Shocked minerals	Yes (9)	Yes (9)	Yes (9)	No	No
Magnetostratigraphy	Yes (10)	Yes (10)	Yes (10)	No	No
Radiometric dates	Yes (11)	Yes (11)	Yes (14)	Yes (11)	Yes (11)
K vertebrates below	Yes	Yes	Yes	Yes	Yes
T vertebrates above	No	No	No	No	No
K leaves below	No	No	No	No	No
T leaves above	Yes (12)	Yes (12)	Yes (12)	No	No

Locality name	(36) Billy Creek	(37) Rick's Place	(38) Iridium Hill	(39) Lerbekmo	(40) Seven Blackfoot
Latitude	47° 32′ 25″ N	47° 31′ 47″ N	47° 31′ 36″ N	47° 36′ 42″ N	47° 26′ 56″ N
Longitude	106° 20′ 43″ W	107° 00′ 27″ W	107° 12′ 41″ W	106° 54′ 42″ W	106° 24′ 48″ W
County	Garfield	Garfield	Garfield	Garfield	Garfield
State/Province	Montana	Montana	Montana	Montana	Montana
Country	USA	USA	USA	USA	USA
Formation(s)	Hell Creek and Fort Union	Hell Creek and Fort Union	Hell Creek and Fort Union	Hell Creek and Fort Union	Hell Creek and Fort Union
Basin	Williston	Williston	Williston	Williston	Williston
Principal reference	Hotton (2002)	Hotton (2002)	Hotton (2002)	Hotton (2002)	Hotton (2002)
Sub decimeter palynology resolution	Yes	Yes	Yes	Yes	Yes
Sampled interval (m)	89	3.7	4	31.7	0.12
K samples for palynology	5	3	3	7	1
T samples for palynology	7	5	2	6	5
Fern-spore spike	Yes	Yes	Yes	Yes	Yes
Iridium anomaly	Yes (amount unspecified)	Yes (amount unspecified)	Yes (amount unspecified)	Yes (amount unspecified)	Yes (amount unspecified)
Claystone layer	Yes	Yes	Yes	Yes	Yes
Shocked minerals	No	No	No	No	No
Magnetostratigraphy	Yes (15)	No	No	No	No
Radiometric dates	Yes (11)	Yes (11)	Yes (11)	Yes (11)	Yes (11)
K vertebrates below	Yes	Yes	Yes	Yes	Yes
T vertebrates above	No	No	No	No	No
K leaves below	No	No	No	No	No
T leaves above	No	No	No	No	No

Locality name	(41) Seven Blackfoot Creek	(42) York Canyon Core[a]	(43) City of Raton[b]	(44) Sugarite	(45) North Ponil
Latitude	47° 31′ 47″ N	36° 52′ 10″ N	36° 54′ 12″ N	36° 56′ 30″ N	36° 36′ 36″ N
Longitude	107° 22′ 38″ W	104° 55′ 30″ W	104° 27′ 10″ W	104° 22′ 30″ W	105° 59′ 00″ W
County	Garfield	Colfax	Colfax	Colfax	Colfax
State/Province	Montana	New Mexico	New Mexico	New Mexico	New Mexico
Country	USA	USA	USA	USA	USA
Formation(s)	Hell Creek and Fort Union	Raton	Raton	Raton	Raton
Basin	Williston	Raton	Raton	Raton	Raton
Principal reference	Tschudy (1970)	Orth et al. (1981)	Orth et al. (1982)	Pillmore et al. (1984)	Pillmore and Fleming (1990)
Sub decimeter palynology resolution	No	Yes	Yes	Yes	Yes
Sampled interval (m)	123	18.5	2.75	2.75	1.5
K samples for palynology	17	11	4	12	4
T samples for palynology	5	11	8	10	5
Fern-spore spike	Yes	Yes	No	Yes (18)	No
Iridium anomaly	No	5.6 ppb	1.0 ppb	2.7 ppb	17.2 ppb
Claystone layer	No	Yes	Yes	Yes	Yes
Shocked minerals	No	No	Yes (17)	Yes (17)	Yes (17)
Magnetostratigraphy	No	Yes (16)	No	No	No
Radiometric dates	Yes (11)	No	No	No	No
K vertebrates below	Yes	No	No	No	No
T vertebrates above	No	No	No	No	No
K leaves below	No	No	No	No	No
T leaves above	No	No	No	No	No

[a] First discovery of K–T boundary in nonmarine rocks.

[b] Also known as Old Raton Pass locality.

Locality name	(46) Dawson North	(47) Crow Creek	(48) Starkville North	(49) Starkville South	(50) Clear Creek North
Latitude	36° 40′ 56″ N	36° 47′ N	37° 07′ 15″ N	37° 06′ 00″ N	37° 05′ 54″ N
Longitude	104° 46′ 15″ W	104° 38′ W	104° 31′ 13″ W	104° 31′ 15″ W	104° 31′ 13″ W
County	Colfax	Colfax	Las Animas	Las Animas	Las Animas
State/Province	New Mexico	New Mexico	Colorado	Colorado	Colorado
Country	USA	USA	USA	USA	USA
Formation(s)	Raton	Raton	Raton	Raton	Raton
Basin	Raton	Raton	Raton	Raton	Raton
Principal reference	Nichols and Pillmore (2000)	Nichols and Pillmore (2000)	Pillmore et al. (1984)	Pillmore et al. (1984)	Pillmore et al. (1984)
Sub decimeter palynology resolution	Yes	Yes	Yes	Yes	Yes
Sampled interval (m)	0.08	2.9	0.4	2.0	0.5
K samples for palynology	4	12	4	8	3
T samples for palynology	7	8	7	8	2
Fern-spore spike	Yes	No	Yes (13)	Yes (19)	No
Iridium anomaly	9.0 ppb	1.2 ppb	6.0 ppb	56.0 ppb	25.0 ppb
Claystone layer	Yes	Yes	Yes	Yes	Yes
Shocked minerals	No	No	Yes (17)	Yes (17)	Yes (17)
Magnetostratigraphy	No	No	No	No	No
Radiometric dates	No	No	No	No	No
K vertebrates below	No	No	No	No	No
T vertebrates above	No	No	No	No	No
K leaves below	No	No	No	No	No
T leaves above	No	No	No	No	No

Locality name	(51) Clear Creek South	(52) Madrid	(53) Berwind Canyon	(54) Long Canyon	(55) Carmel
Latitude	37° 04′ 30″ N	37° 07′ 30″ N	37° 17′ 48″ N	37° 07′ 36″ N	37° 07′ N
Longitude	104° 31′ 01″ W	104° 38′ 00″ W	104° 37′ 13″ W	104° 36′ 00″ W	104° 37′ W
County	Las Animas	Las Animas	Las Animas	Las Animas	Las Animas
State/Province	Colorado	Colorado	Colorado	Colorado	Colorado
Country	USA	USA	USA	USA	USA
Formation(s)	Raton	Raton	Raton	Raton	Raton
Basin	Raton	Raton	Raton	Raton	Raton
Principal reference	Pillmore et al. (1984)	Pillmore et al. (1984)	Orth et al. (1987)	Nichols and Pillmore (2000)	Izett (1990)
Sub decimeter palynology resolution	Yes	Yes	Yes	Yes	Yes
Sampled interval (m)	1.87	0.24	4	0.36	1.35
K samples for palynology	6	4	9	4	0[c]
T samples for palynology	2	4	33	6	0[c]
Fern-spore spike	No	No	Yes (20)	Yes	No
Iridium anomaly	27.0 ppb	8.0 ppb	27.0 ppb	8.2 ppb	13.3 ppb
Claystone layer	Yes	Yes	Yes	Yes	Yes
Shocked minerals	Yes (17)	Yes (17)	Yes (17)	No	Yes
Magnetostratigraphy	No	No	No	No	No
Radiometric dates	No	No	No	No	No
K vertebrates below	No	No	No	No	No
T vertebrates above	No	No	No	No	No
K leaves below	No	No	No	No	No
T leaves above	No	No	No	No	No

[c] No palynological samples collected at this site.

Locality name	(56) South Table Mountain	(57) Kiowa Core	(58) Castle Pines Core	(59) West Bijou Site	(60) Type Lance Area
Latitude	39° 44' 53" N	39° 21' 08.7" N	39° 27' 27" N	39° 34' 14" N	43° 17' 31" N
Longitude	105° 09' 44" W	104° 27' 59.1" W	104° 53' 45" W	104° 18' 09" W	104° 41' 19" W
County	Jefferson	Elbert	Douglas	Arapahoe	Niobrara
State/Province	Colorado	Colorado	Colorado	Colorado	Wyoming
Country	USA	USA	USA	USA	USA
Formation(s)	Denver	Denver	Denver	Denver	Lance and Fort Union
Basin	Denver	Denver	Denver	Denver	Powder River
Principal reference	Kauffman et al. (1990)	Nichols and Fleming (2002)	Nichols and Fleming (2002)	Nichols and Fleming (2002)	Leffingwell (1970)
Sub decimeter palynology resolution	No	Yes	No	Yes	No
Sampled interval (m)	76	522	485	2.3	38
K samples for palynology	2	23	22	5	4
T samples for palynology	3	40	41	10	14
Fern-spore spike	No	Yes	No	Yes	No
Iridium anomaly	No	No	No	6.8 ppb (25)	No
Claystone layer	No	No	No	Yes	No
Shocked minerals	No	Yes	No	Yes	No
Magnetostratigraphy	No	Yes (23)	Yes (23)	Yes	No
Radiometric dates	Yes (21)	Yes (24)	Yes (24)	Yes	No
K vertebrates below	Yes (22)	No	No	Yes (25)	Yes
T vertebrates above	Yes (22)	No	No	Yes (25)	Yes
K leaves below	Yes	No	No	Yes (25, 26)	No
T leaves above	Yes	No	No	Yes (25, 26)	No

Locality name	(61) Dogie Creek	(62) Teapot Dome	(63) Sussex	(64) North Horn Mountain	(65) Red Deer Valley[d]
Latitude	43° 17′ 31″ N	43° 15′ 08″ N	43° 57′ 51″ N	39° 13′ 12″ N	51° 54′ 49″ N
Longitude	104° 41′ 19″ W	106° 03′ 57″ W	106° 20′ 01″ W	111° 11′ 56″ W	112° 56′ 48″ W
County	Niobrara	Converse	Johnson	Emery	Kneehills
State/Province	Wyoming	Wyoming	Wyoming	Utah	Alberta
Country	USA	USA	USA	USA	Canada
Formation(s)	Lance and Fort Union	Lance and Fort Union	Lance and Fort Union	North Horn	Scollard
Basin	Powder River	Powder River	Powder River	(Not given)	Alberta
Principal reference	Bohor et al. (1987a)	Wolfe (1991)	Nichols et al. (1992a)	Difley and Ekdale (2002)	Lerbekmo et al. (1979)
Sub decimeter palynology resolution	Yes	Yes	Yes	No	Yes (10)
Sampled interval (m)	0.65	0.14	1.15	77	1.1
K samples for palynology	9	1	19	(Not given)	7
T samples for palynology	8	14	10	(Not given)	12
Fern-spore spike	Yes	Yes	Yes	No	No
Iridium anomaly	20.8 ppb	22.0 ppb	26.0 ppb	No	3.36 ppb (28)
Claystone layer	Yes	Yes	Yes	No	No
Shocked minerals	Yes	Yes	Yes	No	No
Magnetostratigraphy	No	No	No	No	Yes
Radiometric dates	No	No	No	No	Yes
K vertebrates below	Yes (27)	No	Yes	Yes	Yes
T vertebrates above	Yes (27)	No	No	Yes	No
K leaves below	No	Yes	No	No	No
T leaves above	No	Yes	No	No	No

[d] Also known as Scollard Canyon locality.

Locality name	(66) Coal Valley	(67) Judy Creek 83–313A	(68) Judy Creek 83–368A	(69) Judy Creek 83–401A	(70) Knudsen's Farm
Latitude	53° 02' 40'' N	54° 33' 39'' N	54° 28' 54'' N	54° 31' 22'' N	51° 54' 10'' N
Longitude	116° 42' 40'' W	115° 26' 13'' W	115° 21' 24'' W	115° 26' 10'' W	113° 00' 50'' W
County	(Not given)	(Not given)	(Not given)	(Not given)	Kneehills
State/Province	Alberta	Alberta	Alberta	Alberta	Alberta
Country	Canada	Canada	Canada	Canada	Canada
Formation(s)	Coalspur	Scollard	Scollard	Scollard	Scollard
Basin	Alberta	Alberta	Alberta	Alberta	Alberta
Principal reference	Jerzykiewicz and Sweet (1986)	Sweet et al. (1990)	Sweet and Braman (1992)	Sweet and Braman (1992)	Sweet and Braman (1992)
Sub decimeter palynology resolution	Yes	Yes	Yes	Yes	Yes
Sampled interval (m)	0.8	4.98	2.6	2.8	0.12
K samples for palynology	1	6	7	10	(Not given)
T samples for palynology	8	9	6	9	(Not given)
Fern-spore spike	No	Yes	Yes	Yes	Yes
Iridium anomaly	5.6 ppb (28)	Yes (29; amount unspecified)	No	No	3.4 ppb
Claystone layer	No	No	No	No	Yes
Shocked minerals	No	No	No	No	Yes (30)
Magnetostratigraphy	No	No	No	No	Yes (10)
Radiometric dates	No	No	No	No	Yes (10)
K vertebrates below	No	No	No	No	Yes
T vertebrates above	No	No	No	No	No
K leaves below	No	No	No	No	No
T leaves above	No	No	No	No	No

Locality name	(71) Knudsen's Coulee	(72) Hand Hills	(73) Castle River	(74) Ravenscrag Butte	(75) Frenchman Valley[e]
Latitude	51° 54' 10" N	51° 22' 01" N	49° 32' 30" N	49° 30' 11" N	49° 20' 51" N
Longitude	113° 03' 20" W	112° 11' 21" W	114° 01' 14" W	109° 01' 03" W	108° 25' 10" W
County	Kneehills	Starland	Pincher Creek	White Valley	Val Marie
State/Province	Alberta	Alberta	Alberta	Saskatchewan	Saskatchewan
Country	Canada	Canada	Canada	Canada	Canada
Formation(s)	Scollard	Scollard	Willow Creek	Frenchman and Ravenscrag	Frenchman and Ravenscrag
Basin	Alberta	Alberta	Alberta	Western Canada	Western Canada
Principal reference	Sweet and Braman (1992)	Lerbekmo et al. (1995)	Jerzykiewicz and Sweet (1986)	Lerbekmo (1985)	Lerbekmo et al. (1987)
Sub decimeter palynology resolution	Yes	No	Yes	Yes	Yes
Sampled interval (m)	0.6	(Not given)	(Not given)	2.0	1.0
K samples for palynology	3	2	1	(Not given)	6
T samples for palynology	7	4	3	(Not given)	19
Fern-spore spike	Yes	No	No	No	Yes
Iridium anomaly	12.74 ppb (10)	No	Yes (29; amount unspecified)	Yes (29; amount unspecified)	1.35 ppb (36)
Claystone layer	Yes	Yes (29)	Yes (29)	No	Yes
Shocked minerals	No	No	No	No	No
Magnetostratigraphy	Yes (10)	No	No	Yes	Yes (10)
Radiometric dates	Yes (31, 32)	No	No	No	Yes (32)
K vertebrates below	Yes	No	No	Yes (33, 34)	No
T vertebrates above	No	No	No	Yes (33, 34)	No
K leaves below	No	No	No	No	No
T leaves above	No	No	No	Yes (35)	No

[e] Also called Frenchman River locality.

Locality name	(76) Morgan Creek	(77) Rock Creek East A	(78) Rock Creek West A	(79) Rock Creek West B	(80) Rock Creek West C
Latitude	49° 02′ 45″ N	49° 02′ 09″ N	49° 02′ 40″ N	49° 02′ 20″ N	49° 02′ 36″ N
Longitude	106° 33′ 58″ W	105° 31′ 49″ W	106° 34′ 10″ W	106° 34′ 00″ W	106° 34′ 09″ W
County	Old Post	Old Post	Old Post	Old Post	Old Post
State/Province	Saskatchewan	Saskatchewan	Saskatchewan	Saskatchewan	Saskatchewan
Country	Canada	Canada	Canada	Canada	Canada
Formation(s)	Frenchman and Ravenscrag	Frenchman and Ravenscrag	Frenchman and Ravenscrag	Frenchman and Ravenscrag	Frenchman and Ravenscrag
Basin	Western Canada	Western Canada	Western Canada	Western Canada	Western Canada
Principal reference	Nichols et al. (1986)	Sweet and Braman (1992)	Sweet and Braman (1992)	Sweet and Braman (1992)	Sweet and Braman (2001)
Sub decimeter palynology resolution	Yes	Yes	Yes	Yes	(Not given)
Sampled interval (m)	1.25	1.6	(Not given)	2.55	(Not given)
K samples for palynology	3	5	3	7	(Not given)
T samples for palynology	10	9	3	11	(Not given)
Fern-spore spike	Yes	Yes	Yes	Yes	No
Iridium anomaly	3.0 ppb	Yes (29; amount unspecified)	3.0 ppb	No	No
Claystone layer	Yes	Yes	Yes	Yes	Yes
Shocked minerals	Yes	No	No	No	No
Magnetostratigraphy	Yes (10)	Yes (10)	Yes (10)	Yes (10)	Yes (10)
Radiometric dates	No	No	No	No	No
K vertebrates below	No	No	No	No	No
T vertebrates above	No	No	No	No	No
K leaves below	No	No	No	No	No
T leaves above	No	No	No	No	No

Locality name	(81) Rock Creek West E	(82) Wood Mountain Creek	(83) CCDP Core 13–31–1–2 W3	(84) Police Island	(85) Ugnu SWPT-1 Core
Latitude	49° 02′ 27″ N	49° 25′ 20″ N	49° 05′ 15″ N	64° 52′ 42″ N	70° 24′ 28.2″ N
Longitude	106° 33′ 57″ W	106° 19′ 50″ W	106° 16′ 15″ W	125° 14′ 44″ W	149° 41′ 25.4″ W
County	Old Post	Old Post	Old Post	(Not given)	(Not given)
State/Province	Saskatchewan	Saskatchewan	Saskatchewan	Northwest Territories	Alaska
Country	Canada	Canada	Canada	Canada	USA
Formation(s)	Frenchman and Ravenscrag	Frenchman and Ravenscrag	Frenchman and Ravenscrag	Summit Creek	Lower Ugnu Sands
Basin	Western Canada	Western Canada	Western Canada	Brackett	North Slope
Principal reference	Sweet et al. (1999)	Sweet and Braman (2001)	Sweet et al. (1999)	Sweet et al. (1990)	Frederiksen et al. (1998)
Sub decimeter palynology resolution	Yes	Yes	Yes	Yes	Yes
Sampled interval (m)	0.30	1.0	0.33	32	13.6
K samples for palynology	4	4	5	21	15
T samples for palynology	10	7	9	8	15
Fern-spore spike	Yes	Yes	Yes	No	No
Iridium anomaly	34.7 ppb (10)	Yes (29; amount unspecified)	Yes (29; amount unspecified)	0.3 ppb	No
Claystone layer	Yes	Yes	Yes	No	No
Shocked minerals	No	No	No	No	No
Magnetostratigraphy	Yes (10)	No	No	No	No
Radiometric dates	No	No	No	No	No
K vertebrates below	No	No	No	No	No
T vertebrates above	No	No	No	No	No
K leaves below	No	No	No	No	No
T leaves above	No	No	No	No	No

Locality name	(86) Coll de Nargo	(87) Fontllonga	(88) Campo[f]	(89) Rousset	(90) Albas
Latitude	42° 10′ 60″ N	41° 58′ 00″ N	42° 23′ 00″ N	44° 28′ 60″ N	44° 28′ 0″ N
Longitude	01° 19′ 0″ E	0° 51′ 00″ E	0° 24′ 00″ E	06° 15′ 0″ E	01° 13′ 60″ E
County	(Not given)	(Not given)	(Not given)	(Not given)	(Not given)
State/Province	Catalonia	Lleida, Catalonia	Huesca	Alsace	Alsace
Country	Spain	Spain	Spain	France	France
Formation(s)	Tremp	Tremp	Tremp	(Not given)	(Not given)
Basin	Tremp	Tremp	Tremp	(Not given)	(Not given)
Principal reference	Ashraf and Erben (1986)	Médus et al. (1992)	Fernández-Marrón et al. (2004)	Ashraf and Erben (1986)	Ashraf and Erben (1986)
Sub decimeter palynology resolution	No	No	No	No	No
Sampled interval (m)	40	225	80	26	17
K samples for palynology	27	3	1	7	1
T samples for palynology	4	7	5	26	3
Fern-spore spike	No	No	No	No	No
Iridium anomaly	No	No	No	No	No
Claystone layer	No	No	No	No	No
Shocked minerals	No	Yes (37)	No	No	No
Magnetostratigraphy	No	Yes (38)	Yes	No	No
Radiometric dates	No	No	No	No	No
K vertebrates below	No	Yes (39)	Yes	Yes	No
T vertebrates above	No	Yes (39)	No	No	No
K leaves below	No	No	No	No	No
T leaves above	No	Yes	No	No	No

[f]Marine section for which terrestrial palynomorph data are available.

Locality name	(91) Geulhemmerberg[g]	(92) Curfs Quarry[g]	(93) Kawaruppu[g]	(94) Nanxiong	(95) Baishantou
Latitude	50° 52' 00'' N	50° 52' 00'' N	43° 04' 00'' N	25° 05' 48'' N	49° 10' N
Longitude	5° 46' 60'' E	5° 46' 60'' E	143° 41' 00'' E	114° 16' 30'' E	129° 30' E
County	(Not given)	(Not given)	(Not given)	Nanxiong	(Not given)
State/Province	Limburg	Limburg	Hokkaido	Guangdong	Heilongjiang
Country	Netherlands	Netherlands	Japan	China	China
Formation(s)	Maastricht	Maastricht	Katsuhira	Nanxiong and Shanghu	Wuyun
Basin	(Not given)	(Not given)	(Not given)	Nanxiong	(Not given)
Principal reference	Brinkhuis and Schiøler (1996)	Herngreen et al. (1998)	Saito et al. (1986)	Taylor et al. (2006)	Sun et al. (2002)
Sub decimeter palynology resolution	Yes	No	Yes	No	No
Sampled interval (m)	1.25	27	0.85	300	15
K samples for palynology	3	21	4	(40; unspecified among 425)	(Not given)
T samples for palynology	21	13	13	(40; unspecified among 425)	(Not given)
Fern-spore spike	No	No	Yes	No	No
Iridium anomaly	No	No	No	No[h]	No
Claystone layer	No	No	Yes	No	No
Shocked minerals	No	No	No	No	No
Magnetostratigraphy	No	No	No	Yes (41)	No
Radiometric dates	No	No	No	Yes (42)	No
K vertebrates below	No	No	No	Yes	Yes
T vertebrates above	No	No	No	Yes	No
K leaves below	No	No	No	No	No
T leaves above	No	No	No	No	Yes

[g] Marine section for which terrestrial palynomorph data are available.

[h] Iridium in dinosaur eggshell fragments reported at several levels, but not in a claystone layer.

Locality name	(96) Blagoveshchensk	(97) Kundur	(98) Beringovskoe	(99) Sinegorsk	(100) El Kef[a]
Latitude	50° 16' N	49° 6' N	63° 02' N	47° 10' N	36° 10' 56'' N
Longitude	127° 31' E	130° 45' E	179° 18' E	142° 30' E	08° 42' 53'' E
County	(Not given)	(Not given)	(Not given)	(Not given)	(Not given)
State/Province	Amurskaya Oblast'	Amurskaya Oblast'	Chukotsky Okrug	Sakhalinskaya Oblast'	(Not given)
Country	Russia	Russia	Russia	Russia	Tunisia
Formation(s)	Tsagayan	Tsagayan	Koryak	Krasnoyarka	El Haria
Basin	Zeya–Bureya	Zeya–Bureya	Kakanaut River	(Not given)	(Not given)
Principal reference	Markevich and Bugdaeva (2001)	Markevich and Bugdaeva (2001)	Markevich and Bugdaeva (2001)	Markevich and Bugdaeva (2001)	Méon (1990)
Sub decimeter palynology resolution	No	No	No	No	Yes
Sampled interval (m)	(Not given)	(Not given)	(Not given)	(Not given)	450
K samples for palynology	(Not given)	(Not given)	(Not given)	(Not given)	60
T samples for palynology	(Not given)	(Not given)	(Not given)	(Not given)	19
Fern-spore spike	No	No	No	No	5 ppb (43) – estimated
Iridium anomaly	No	No	No	No	No
Claystone layer	No	No	No	No	No
Shocked minerals	No	No	No	No	No
Magnetostratigraphy	No	No	No	No	No
Radiometric dates	No	No	No	No	No
K vertebrates below	Yes	Yes	No	Yes	No
T vertebrates above	No	No	No	No	No
K leaves below	No	Yes	Yes	No	No
T leaves above	No	Yes	Yes	No	No

[a] Marine section for which terrestrial palynomorph data are available.

Locality name	(101) Anjar	(102) Seymour Island[i]	(103) Moody Creek Mine	(104) Compressor Creek	(105) Mid-Waipara River[i]
Latitude	23° 36' N	64° 30' S	42° 23' 18" S	42° 22' 31" S	43° 03' 44" S
Longitude	70° 12' E	56° 45' W	171° 16' 40" E	171° 18' 35" E	172° 34' 56" E
County	(Not given)	(Not given)	(Not given)	(Not given)	(Not given)
State/Province	Gujarat, Kutch	(Not given)	South Island	South Island	South Island
Country	India	Antarctica	New Zealand	New Zealand	New Zealand
Formation(s)	Deccan intertrappeans	López de Bertodano	Paparoa Coal Measures	Paparoa Coal Measures	Conway
Basin	(Not given)	(Not given)	Greymouth Coalfield	Greymouth Coalfield	(Not given)
Principal reference	Bhandari et al.(1995)	Askin (1990)	Vajda et al. (2001)	Vajda et al. (2004)	Vajda et al. (2001)
Sub decimeter palynology resolution	No	No	Yes	Yes	Yes
Sampled interval (m)	(Not given)	210	0.8	3.9	1.85
K samples for palynology	(Not given)	7	6	11	17
T samples for palynology	(Not given)	8	22	14	24
Fern-spore spike	No	No	Yes	Yes	Yes
Iridium anomaly	1.27 ppb	1.6 ppb (44)	71 ppb	No	0.49 ppb
Claystone layer	No	No	No	No	No
Shocked minerals	No	No	No	No	No
Magnetostratigraphy	No	No	No	No	No
Radiometric dates	Yes	No	No	No	No
K vertebrates below	Yes	No	No	No	Yes (45)
T vertebrates above	No	No	No	No	No
K leaves below	No	No	Yes	Yes	Yes
T leaves above	No	No	Yes	Yes	Yes

[i] Marine section for which terrestrial palynomorph data are available.

(1) Hicks *et al.* (2002)

(2) Pearson *et al.* (2001)

(3) Pearson *et al.* (2002)

(4) Hunter (1999)

(5) Johnson (1992)

(6) Johnson (2002)

(7) Lerbekmo and Coulter (1984)

(8) Hunter *et al.* (1997)

(9) Bohor *et al.* (1984)

(10) Lerbekmo *et al.* (1996)

(11) Swisher *et al.* (1993)

(12) Johnson and Hickey (1990)

(13) Tschudy *et al.* (1984)

(14) Baadsgaard and Lerbekmo (1980)

(15) Archibald *et al.* (1982)

(16) Shoemaker *et al.* (1987)

(17) Izett (1990)

(18) Nichols and Fleming (1990)

(19) Pillmore *et al.* (1998)

(20) Pillmore *et al.* (1988)

(21) Newman (1979)

(22) Brown (1943)

(23) Hicks *et al.* (2003)

(24) Raynolds and Johnson (2003)

(25) Barclay *et al.* (2003)

(26) Johnson *et al.* (2003)

(27) Leffingwell (1970)

(28) Lerbekmo *et al.* (1987)

(29) Sweet and Braman (2001)

(30) Bohor *et al.* (1987a)

(31) Lerbekmo *et al.* (1979)

(32) Baadsgaard *et al.* (1988)

(33) Fox (1989)

(34) Fox (2002)

(35) McIver and Basinger (1993)

(36) Sweet and Braman (1992)

(37) Galbrun *et al.* (1993)

(38) López-Martínez *et al.* (1998)

(39) López-Martínez *et al.* (1999)

(40) Stets *et al.* (1996)

(41) Zhao *et al.* (1991)

(42) Rigby *et al.* (1993)

(43) Robin *et al.* (1991)

(44) Elliot *et al.* (1994)

(45) Wilson *et al.* (2005)

References

Akhmetiev, M. (2004). Biosphere crisis at the Cretaceous-Paleogene boundary. In *Proceedings of the 3rd Symposium on Cretaceous Biota and the K/T Boundary in Heilongjiang River Area*, ed. G. Sun, Y. W. Sun, M. Akhmetiev, and A. R. Ashraf, pp. 7-16. Changchun, China: Jilin University Research Center of Palaeontology and Stratigraphy.

Albritton, C. C., Jr. (1989). *Catastrophic Episodes in Earth History*. London: Chapman and Hall Ltd.

Allaby, M. and Lovelock, J. (1983). *The Great Extinction*. Garden City, NY: Doubleday.

Alt, D., Sears, J. M., and Hyndman, D. W. (1988). Terrestrial maria: the origins of large basaltic plateaus, hotspot tracks and spreading ridges. *Journal of Geology* **96**, 647-62.

Alvarez, L. W., Alvarez, W., Asaro, F., and Michel, H. V. (1980). Extraterrestrial cause for the Cretaceous-Tertiary extinction. *Science* **208**, 1095-1108.

Alvarez, W. (1997). *T. Rex and the Crater of Doom*. Princeton, NJ: Princeton University Press.

Alvarez, W., Alvarez, L. W., Asaro, F., and Michel, H. V. (1984). The end of the Cretaceous: sharp boundary or gradual transition? *Science* **223**, 1183-6.

Archibald, J. D. (1996). *Dinosaur Extinction and the End of an Era: What the Fossils Say*. New York, NY: Columbia University Press.

Archibald, J. D., Butler, R. F., Lindsay, E. H., Clemens, W. A., and Dingus, L. (1982). Upper Cretaceous-Paleocene biostratigraphy and magnetostratigraphy, Hell Creek and Tullock Formations, northeastern Montana. *Geology* **10**, 153-9.

Arens, N. C. and Jahren, A. H. (2000). Carbon isotopic excursion in atmospheric CO_2 at the Cretaceous-Tertiary boundary: evidence from terrestrial sediments. *Palaios* **15**, 314-22.

Arens, N. C. and Jahren, A. H. (2002). Chemostratigraphic correlation of four fossil-bearing sections in southwestern North Dakota. In *The Hell Creek Formation and the Cretaceous-Tertiary Boundary in the Northern Great Plains: An Integrated Continental Record of the End of the Cretaceous*, ed. J. H. Hartman, K. R. Johnson, and D. J. Nichols. *Geological Society of America Special Paper* **361**, 75-93.

Ashraf, A. R. and Erben, H. K. (1986). Palynologische Untersuchungen an der Kreide/ Tertiär-Grenze west-Mediterraner Regionen. *Palaeontographica B* **200**, 111-63.

Ashraf, A. R. and Stinnesbeck, W. (1988). Pollen und Sporen an der Kreide-Tertiärgrenze im Staate Pernambuco, NE Brasilien. *Palaeontographica B* **208**, 39-51.

Askin, R. A. (1988a). Campanian to Paleocene palynological succession of Seymour and adjacent islands, northeastern Antarctic Peninsula. In *Geology and Paleontology of Seymour Island, Antarctic Peninsula*, ed. R. M. Feldman and M. O. Woodburne. *Geological Society of America Memoir* **169**, 131-53.

Askin, R. A. (1988b). The palynological record across the Cretaceous/Tertiary transition on Seymour Island, Antarctica. In *Geology and Paleontology of Seymour Island, Antarctic Peninsula*, ed. R. M. Feldman and M. O. Woodburne. *Geological Society of America Memoir* **169**, 155-62.

Askin, R. A. (1990). Campanian to Paleocene spore and pollen assemblages of Seymour Island, Antarctica. In *Proceedings of the 7th International Palynological Congress*, ed. E. M. Truswell and J. A. K. Owen. *Review of Palaeobotany and Palynology* **65**, 105-13.

Axelrod, D. I. and Raven, P. H. (1978). Late Cretaceous and Tertiary vegetation history of Africa. In *Biogeography and Ecology of Southern Africa*, ed. M. J. A. Werger, pp. 77-130. The Hague: Dr W. Junk.

Baadsgaard, H. and Lerbekmo, J. F. (1980). A Rb/Sr age for the Cretaceous Tertiary boundary (Z coal), Hell Creek, Montana. *Canadian Journal of Earth Sciences* **17**, 671-3.

Baadsgaard, H., Lerbekmo, J. F., and McDougall, I. (1988). A radiometric age for the Cretaceous–Tertiary boundary based upon K-Ar, Rb-Sr, and U-Pb ages of bentonites from Alberta, Saskatchewan, and Montana. *Canadian Journal of Earth Sciences* **25**, 1088-97.

Baksi, S. and Deb, U. (1981). Palynology of the Upper Cretaceous of the Bengal Basin, India. *Review of Palaeobotany and Palynology* **31**, 335-65.

Barclay, R. S., Johnson, K. R., Betterton, W. J., and Dilcher, D. L. (2003). Stratigraphy and megaflora of a K-T boundary section in the eastern Denver Basin, Colorado. *Rocky Mountain Geology* **38**, 45-71.

Batten, D. J. (1981). Stratigraphic, palaeogeographic and evolutionary significance of Late Cretaceous and early Tertiary Normapolles pollen. *Review of Palaeobotany and Palynology* **35**, 125-37.

Batten, D. J. (1984). Palynology, climate and the development of Late Cretaceous floral provinces in the Northern Hemisphere; a review. In *Fossils and Climate*, ed. P. Brenchly, pp. 127-64. Chichester: Wiley.

Beerling, D. J., Lomax, B. H., Upchurch, G. R., Jr., *et al.* (2001). Evidence for the recovery of terrestrial ecosystems ahead of marine primary production following a biotic crisis at the Cretaceous–Tertiary boundary. *Journal of the Geological Society, London* **158**, 737-40.

Belcher, C. M., Collinson, M. E., Sweet, A. R., Hildebrand, A. R., and Scott, A. C. (2003). Fireball passes and nothing burns - the role of thermal radiation in the Cretaceous–Tertiary event: evidence from the charcoal record of North America. *Geology* **31**, 1061-4.

Bhandari, N., Shukla, P. N., Ghevariya, Z. G., and Sundaram, S. M. (1995). Impact did not trigger Deccan volcanism: evidence from Anjar K/T boundary intertrappean sediments. *Geophysical Research Letters* **22**, 433-6.

Bhandari, N., Shukla, P. N., Ghevariya, Z. G., and Sundaram, S. M. (1996). K/T boundary layer in Deccan Intertrappeans at Anjar, Kutch. In *The Cretaceous-Tertiary Event and Other Catastrophes in Earth History*, ed. G. Ryder, D. Fastovsky, and S. Gartner. *Geological Society of America Special Paper* **307**, 417-24.

Bohor, B. F., Foord, E. E., Modreski, P. J., and Triplehorn, D. M. (1984). Mineralogic evidence for an extraterrestrial impact event at the Cretaceous-Tertiary boundary. *Science* **224**, 867-9.

Bohor, B. F., Modreski, P. J., and Foord, E. E. (1987a). Shocked quartz in the Cretaceous-Tertiary boundary clays: evidence for a global distribution. *Science* **236**, 705-9.

Bohor, B. F., Triplehorn, D. M., Nichols, D. J., and Millard, H. T., Jr. (1987b). Dinosaurs, spherules and the "magic layer": a new K-T boundary clay site in Wyoming. *Geology* **15**, 896-9.

Braman, D. R. and Sweet, A. R. (1999). Terrestrial palynomorph biostratigraphy of the Cypress Hills, Wood Mountain, and Turtle Mountain areas (Upper Cretaceous-Paleocene) of western Canada. *Canadian Journal of Earth Sciences* **36**, 725-41.

Bratseva, G. M. (1969). Palinologicheskiye issledovaniya verkhnego mela I paleogena Dal'nego Vostoka [Palynological investigation of Upper Cretaceous and Paleogene of the Far East]. *Trudy Geologicheskiy Institut, Akademiya Nauk SSSR* [*Transactions of the Geological Institute, USSR Academy of Sciences*] **207**, 5-56, 64 pls. [in Russian]

Brett, R. (1992). The Cretaceous-Tertiary extinction: a lethal mechansim involving anhydrite target rocks. *Geochimica et Cosmochimica Acta* **56**, 3603-6.

Brinkhuis, H. and Schiøler, P. (1996). Palynology of the Geulhemmerberg Cretaceous/Tertiary boundary section (Limburg, SE Netherlands). *Geologie en Mijnbouw* **75**, 193-213.

Brown, R. W. (1939). Fossil plants from the Colgate Member of the Fox Hills Sandstone and adjacent strata. *US Geological Survey Professional Paper* **189-I**, 239-75.

Brown, R. W. (1943). The Cretaceous-Tertiary boundary in the Denver Basin, Colorado. *Geological Society of America Bulletin* **54**, 65-86.

Brown, R. W. (1962). Paleocene flora of the Rocky Mountains and Great Plains. *US Geological Survey Professional Paper* **375**.

Buck, B. J., Hanson, A. D., Hengst, R. A., and Hu, S-S. (2004). "Tertiary dinosaurs" in the Nanxiong Basin, southern China, are reworked from the Cretaceous. *Journal of Geology* **112**, 111-18.

Bugdaeva, E. V., ed. (2001). *Flora and Dinosaurs at the Cretaceous-Paleogene Boundary of Zeya-Bureya Basin*. Vladivostok, Russia: Dalnauka. [in Russian]

Burnham, R. J. and Wing, S. L. (1989). Temperate forest litter accurately reflects stand composition and structure. *American Journal of Botany* **76**, 159.

Campbell, I. D. (1999). Quaternary pollen taphonomy: examples of differential redeposition and differential preservation. *Palaeogeography, Palaeoclimatology, Palaeoecology* **149**, 245–56.

Cande, S. C. and Kent, D. V. (1995). Revised calibration of the geomagnetic polarity timescale for the Late Cretaceous and Cenozoic. *Journal of Geophysical Research* **100**, 6093–5.

Carlisle, D. B. (1995). *Dinosaurs, Diamonds, and Things from Outer Space: the Great Extinction.* Stanford, CA: Stanford University Press.

Carlisle, D. B. and Braman, D. R. (1991). Nanometre-size diamonds in the Cretaceous/ Tertiary boundary clay of Alberta. *Nature*, **352**, 708–9.

Chen, Pei-ji (1983). A survey of the non-marine Cretaceous in China. *Cretaceous Research* **4**, 123–43.

Chen, Pei-ji (1996). Freshwater biota, stratigraphic correlation of Late Cretaceous of China. *Geological Society of India Memoir* **37**, 35–62.

Chen, Pei-ji, Li, J., Matsukawa, M., *et al.* (2006). Geological ages of dinosaur-track-bearing formations in China. *Cretaceous Research* **27**, 22–32.

Chlonova, A. F. (1962). Some morphological types of spores and pollen grains from Upper Cretaceous of eastern part of West Siberian Lowland. *Pollen et Spores* **4**, 297–309.

Cojan, I. (1989). Discontinuités majeures en mileu continental. Proposition de corrélation avec des événements globaux (Bassin de Provence, S. France, Passage Crétacé/Tertiaire). *Comptes Rendus de l'Académie des Sciences, II* **309**, 1013–18.

Couper, R. A. (1960). New Zealand Mesozoic and Cainozoic plant microfossils. *New Zealand Geological Survey Paleontological Bulletin* **32**.

Courtillot, V., Feraud, G., Maluski, H., *et al.* (1988). Deccan flood basalts and Cretaceous/Tertiary boundary. *Nature* **333**, 843–6.

Courtillot, V., Vandamme, D., and Besse, J. (1990). Deccan volcanism at the Cretaceous–Tertiary boundary; data and inferences. In *Global Catastrophes in Earth History; An Interdisciplinary Conference on Impacts, Volcanism, and Mass Mortality,* ed. V. L. Sharpton, and P. D. Ward. *Geological Society of America Special Paper* **247**, 401–9.

Courtillot, V., Gallet, Y., Rocchia, R., *et al.* (2000). Cosmic markers, [40]Ar/[39]Ar dating and paleomagnetism of the K/T sections in the Anjar area of the Deccan Large Igneous Province. *Earth and Planerary Science Letters* **182**, 137–56.

Covey, C., Thompson, S. L., Weissman, P. R., and MacCracken, M. C. (1994). Global climatic effects of atmospheric dust from an asteroid or comet impact on Earth. *Global Planetary Change* **9**, 263–73.

Crane, P. R. and Lidgard, S. (1989). Angiosperm diversification and paleolatitudinal gradients in Cretaceous floristic diversity. *Science* **246**, 675–8.

Cronquist, A. (1981). *An integrated system of classification of flowering plants.* New York, NY: Columbia University Press.

Desor, E. (1847). Sur la terrain danien, nouvel etage de la Craie. *Bulletin de la Société Géologique de France, serie 2,* **4**, 179–82.

D'Hondt, S., Pilson, M. E. Q., Sigurdsson, H., Hanson, A. K., and Carey, S. (1994). Surface-water acidification and extinction at the Cretaceous–Tertiary boundary. *Geology* **22**, 983–6.

D'Hondt, S., Herbert, T. D., King, J., and Gibson, C. (1996). Planktic foraminifera, asteroids, and marine production: death and recovery at the Cretaceous–Tertiary boundary. In *The Cretaceous–Tertiary Event and Other Catastrophes in Earth History*, ed. G. Ryder, D. Fastovsky, and S. Gartner. *Geological Society of America Special Paper* **307**, 303–17.

Difley, R. and Ekdale, A. A. (1999). Biostratigraphic aspects of the Cretaceous–Tertiary (KT) boundary interval at North Horn Mountain, Emery County, Utah. In *Vertebrate Paleontology in Utah*, ed. D. D. Gillette. *Utah Geological Survey Miscellaneous Publication* **99–1**, 389–98.

Difley, R. and Ekdale, A. A. (2002). Faunal implications of an environmental change before the Cretaceous–Tertiary (K–T) transition in central Utah. *Cretaceous Research* **23**, 315–31.

DiMichele, W. A., Phillips, T. L., and Olmstead, R. G. (1987). Opportunistic evolution: abiotic environmental stress and the fossil record of plants. *Review of Palaeobotany and Palynology* **50**, 151–78.

Doerenkamp, A., Jardine, S., and Moreau, P. (1976). Cretaceous and Tertiary palynomorph assemblages from Banks Island and adjacent areas (N. W. T.). *Bulletin of Canadian Petroleum Geology* **24**, 372–417.

Dogra, N. N., Singh, R. Y., and Kulshreshtha, S. K. (1994). Palynostratigraphy of infra-trappean Jabalpur and Lameta formations (Lower and Upper Cretaceous) in Madhya Pradesh, India. *Cretaceous Research* **15**, 205–15.

Dogra, N. N., Singh, Y. R., and Singh, R. Y. (2004). Palynological age from the Anjar intertrappeans, Kutch district, Gujarat: age implications. *Current Science* **86**, 1596–7.

Doher, L. I. (1980). Palynomorph preparation procedures currently used in the Paleontology and Stratigraphy Laboratories, US Geological Survey. *US Geological Survey Circular* **830**.

Dorf, E. (1940). Relationship between floras of the type Lance and Fort Union Formations. *Geological Society of America Bulletin* **51**, 213–36.

Dorf, E. (1942a). Application of paleobotany to the Cretaceous–Tertiary boundary problem. *Transactions of the New York Academy of Sciences, Ser. II* **4**, 73–8.

Dorf, E. (1942b). Upper Cretaceous Floras of the Rocky Mountain Region. Part 2. Flora of the Lance Formation at its type locality, Niobrara County, Wyoming. *Carnegie Institute of Washington Publication* **508**.

Drinnan, A. N. and Crane, P. R. (1990). Cretaceous paleobotany and its bearing on the biogeography of austral angiosperms. In *Antarctic Paleobiology. Its Role in the Reconstruction of Gondwana*, ed. T. N. Taylor and E. L. Taylor, pp. 192–219. New York, NY: Springer-Verlag.

Dutra, T. L. and Batten, D. J. (2000). Upper Cretaceous floras of King George Island, west Antarctica, and their palaeoenvironmental and phytogeographic implications. *Cretaceous Research* **21**, 181–209.

Elliot, D. H., Askin, R. A., Kyte, F. T., and Zinsmeister, W. J. (1994). Iridium and dinocysts at the Cretaceous–Tertiary boundary on Seymour Island, Antarctica: implications for the K–T event. *Geology* **22**, 675–8.

Ellis, B., Johnson, K. R., and Dunn, R. E. (2003). Evidence for an *in situ* early Paleocene rainforest from Castle Rock, Colorado. *Rocky Mountain Geology* **38**, 73–100.

Erben, H. K., Ashraf, A. R., Böhm, H., *et al.* (1995). Die Kreide/Tertiär-Grenze im Nanxiong-Becken (Kontinentalfazies, Südostchina). *Erdwissenschaftliche Forschung* **32**. [in German with English abstract]

Erdtman, G. (1957). *Pollen and Spore Morphology/Plant Taxonomy – Gymnospermae, Pteridophyta, Bryophyta (Illustrations).* Stockholm: Almqvist & Wiksell.

Erdtman, G. (1965). *Pollen and Spore Morphology/Plant Taxonomy – Gymnospermae, Bryophyta.* Stockholm: Almqvist & Wiksell.

Erdtman, G. (1966). *Pollen Morphology and Plant Taxonomy – Angiosperms.* New York, NY: Hafner.

Fernández-Marrón, M. T., López-Martinez, N., Fonolla, J. F., and Valle-Hernández, M. F. (2004). The palynological record across the Cretaceous–Tertiary boundary in differing palaeogeographical settings from the southern Pyrenees, Spain. In *The Palynology and Micropalaeontology of Boundaries*, ed. A. B. Beaudoin and M. J. Head. *Geological Society, London, Special Publication* **230**, 243–55.

Fleming, R. F. and Nichols, D. J. (1990). The fern-spore abundance anomaly at the Cretaceous–Tertiary boundary – a regional bioevent in western North America. In *Extinction Events in Earth History*, ed. E. G. Kauffman and O. H. Walliser. *Lecture Notes in Earth Sciences* **30**, 347–9.

Fouch, T. D., Hanley, J. H., Forester, R. M. *et al.* (1987). Chart showing lithology, mineralogy and paleontology of the nonmarine North Horn Formation and Flagstaff Member of the Green River Formation, Price Canyon, central Utah; a principal reference section: *US Geological Survey Miscellaneous Investigations Map* **I-1797A**, 1 chart.

Fox, R. C. (1989). The Wounded Knee local fauna and mammalian evolution near the Cretaceous–Tertiary boundary, Saskatchewan, Canada. *Palaeontographica, Abt. A* **208**, 11–59.

Fox, R. C. (2002). The oldest Cenozoic mammal? *Journal of Vertebrate Paleontology* **22**, 456–9.

Fox, S. K., Jr. and Ross, R. J., Jr. (1942). Foraminiferal evidence for the Midway (Paleocene) Age of the Cannonball Formation in North Dakota. *Journal of Paleontology* **16**, 660–73.

Frankel, C. (1999). *The End of the Dinosaurs.* Cambridge: Cambridge University Press.

Frederiksen, N. O. (1989). Changes in floral diversities, floral turnover rates, and climates in Campanian and Maastrichtian time, North Slope of Alaska. *Cretaceous Research* **10**, 249–66.

Frederiksen, N. O., Ager, T. A., and Edwards, L. E. (1988). Palynology of Maastrichtian and Paleocene rocks, lower Colville River region, North Slope of Alaska. *Canadian Journal of Earth Sciences* **25**, 512–27.

Frederiksen, N. O., Anderle, V. A. S., Sheehan, T. P., *et al.* (1998). Palynological dating of Upper Cretaceous to middle Eocene strata in the Sagavanirktok and Canning Formations, North Slope of Alaska. *US Geological Survey Open-File-Report* **98-471**.

Funkhouser, J. W. (1961). Pollen of the genus *Aquilapollenites*. *Micropaleontology* **7**, 193–8.

Galbrun, B., Feist, M., Colombo, F., Rocchia, R., and Tambareau, Y. (1993). Magnetostratigraphy and biostratigraphy of Cretaceous–Tertiary continental deposits, Ager Basin, Province of Lerida, Spain. *Palaeogeography, Palaeoclimatology, Palaeoecology* **102**, 41–52.

Gardner, A. F. and Gilmour, I. (2002). Organic geochemical investigation of terrestrial Cretaceous–Tertiary boundary successions from Brownie Butte, Montana, and the Raton Basin, New Mexico. In *Catastrophic Events and Mass Extinctions: Impacts and Beyond*, ed. C. Koeberl and K. G. McLeod. *Geological Society of America Special Paper* **356**, 351–62.

Gemmill, C. E. C. and Johnson, K. R. (1997). Paleoecology of a late Paleocene (Tiffanian) megaflora from the northern Great Divide Basin, Wyoming. *Palaios* **12**, 439–48.

Geological Institute, Russian Academy of Sciences (2003). *The 2nd International Symposium on Cretaceous Biota and K/T Boundary in Amur (Heilongjiang) River Area, III. Field Excursion Guidebook*. Moscow: ГЕОС [Geos].

Germeraad, J. H., Hopping, C. A., and Muller, J. (1968). Palynology of Tertiary sediments from tropical areas. *Review of Palaeobotany and Palynology* **6**, 189–348.

Gilmore, C. W. (1946). Reptilian fauna of the North Horn Formation of central Utah. *U.S. Geological Survey Professional Paper* **210-C**, 29–53.

Gilmour, I., Russell, S. S., Pillinger, C. T., Lee, M., and Arden, J. W. (1992). Origin of microdiamonds in KT boundary clays. *Abstracts of the Lunar and Planetary Science Conference* **23**, 413.

Godefroit, P., Zan, S., and Jin, L. (2001). The Maastrichtian (Late Cretaceous) lambeosaurine dinosaur *Charnosaurus jiayinensis* from north-eastern China. *Bulletin de l'Instutut Royal des Sciences Naturelles de Belgique, Sciences de la Terre* **71**, 119–68.

Godefroit, P., Bolotsky, Y., and Alifanov, V. (2003). A remarkable hollow-crested hadrosaur from Russia: an Asian origin for lambeosaurines. *Comptes Rendus Palevol* **2**, 143–51.

Golovneva, L. B. (1994a). The flora of the Maastrichtian–Danian deposits of the Koryak Uplands, northeast Russia. *Cretaceous Research* **15**, 89–100.

Golovneva, L. B. (1994b) *Maastrichtian–Danian floras of Koryak Upland*. St. Petersburg, Russia: Botanical Institute, Russian Academy of Science. [in Russian]

Golovneva, L. B. (1995). Environmental changes and patterns of floral evolution during the Cretaceous–Tertiary transition in northeastern Asia. *Paleontological Journal* **29**, 36–49.

Golovneva, L., Bugdaeva, E., Sun, G., Akhmetiev, M., and Kodrul, T. (2004). Systematic composition and age of floristic assemblages from the Kundur and Taipinglinchang formations. In *Proceedings of the 3rd Symposium on Cretaceous Biota and the K/T Boundary in Heilongjiang River Area*, ed. G. Sun, Y. W. Sun, M. Akhmetiev, and A. R. Ashraf, pp. 23–6. Changchun, China: Jilin University.

Gradstein, F., Ogg, J., and Smith, A. (2004). *A Geologic Time Scale 2004*. Cambridge: Cambridge University Press.

Greuter, W., McNeill, J., Barrie, F. R., Burdet, H. M. *et al.*, eds. (2000). *International Code of Botanical Nomenclature (Saint Louis Code)*. Königstein: Koeltz Scientific Books.

Hansen, H. J., Mohabey, D. M., and Toft, P. (2001). No K/T boundary at Anjar, Gujarat, India: evidence from magnetic susceptibility and carbon isotopes. *Proceedings of the Indian Academy of Science (Earth and Planetary Sciences)* **110**, 133–42.

Hao, Y. and Guan, S. (1984). The Lower-Upper Cretaceous and Cretaceous-Tertiary boundaries in China. *Bulletin of the Geological Society of Denmark* **33**, 129-38.

Hao, Y., Yu, J., Guan, S., and Sun, M. (1979). Some Late Cretaceous and early Tertiary assemblages of Ostracoda, spores and pollen in China. In *Cretaceous-Tertiary Boundary Events, Symposium II, Proceedings*, ed. W. K. Christensen and T. Birkelund. *University of Copenhagen Geological Museum, Contributions to Palaeontology* **294**, 251-5.

Haq, B. U., Hardenbol, J., and Vail, P. R. (1988). Mesozoic and Cenozoic chronostratigraphy and cycles of sea-level change. In *Sea-Level Changes: An Integrated Approach*, ed. C. A. Ross, J. Van Wagoner, and C. K. Wilgus. *Society of Economic Paleontologists and Mineralogists Special Publication* **42**, 71-108.

Harland, W. B., Cox, A. V., Llewellyn, P. G., *et al.* (1982). *A Geologic Time Scale*. Cambridge: Cambridge University Press.

Harland, W. B., Armstrong, R. L., Cox, A. V., *et al.* (1989). *A Geologic Time Scale 1989*. Cambridge: Cambridge University Press.

Harries, P. J., Kauffman, E. G., and Hansen, T. A. (1996). Models for biotic survival following mass extinction. In *Biotic Recovery from Mass Extinction Events*, ed. M. B. Hart. *Geological Society Special Publication* **102**, 41-60.

Hartman, J. H., Johnson, K. R., and Nichols, D. J., eds. (2002). *The Hell Creek Formation and the Cretaceous-Tertiary Boundary in the Northern Great Plains: An Integrated Continental Record of the End of the Cretaceous*. Geological Society of America Special Paper 361.

Heer, O. (1882). Die Fossile Flora Grönlands, 1. *Flora Fossilis Arctica* **6**, 1-147.

Heer, O. (1883). Die Fossile Flora Grönlands, 2. *Flora Fossilis Arctica* **7**, 148-275.

Herman, A. B. and Spicer, R. A. (1995). Latest Cretaceous flora of northeastern Russia and the "Terminal Cretaceous Event" in the Arctic. *Paleontological Journal* **29**, 22-35.

Herman, A. B. and Spicer, R. A. (1997). The Koryak flora: did the early Tertiary deciduous flora begin in the late Maastrichtian of northeastern Russia? *Mededelingen Nederlands Instituut voor Toegepaste Geowetenschappen TNO* **58**, 87-92.

Herngreen, G. F. W. and Chlonova, A. F. (1981). Cretaceous microfloral provinces. *Pollen et Spores* **23**, 441-555.

Herngreen, G. F. W., Kedves, M., Rovnina, L. V., and Smirnova, S. B. (1996). Cretaceous palynofloral provinces: a review. In *Palynology: Principles and Applications*, vol. 3, ed. J. Jansonius and D. C. McGregor, pp. 1157-88. Dallas, TX: American Association of Stratigraphic Palynologists.

Herngreen, G. F. W., Schuurman, H. A. H. M., Verbeek, J. W., *et al.* (1998). Biostratigraphy of Cretaceous/Tertiary boundary strata in the Curfs quarry, the Netherlands. *Mededlingen Nederlands Instituut voor Toegepaste Geowetenschappen TNO* **61**.

Hickey, L. J. (1973). Classification of the architecture of dicotyledonous leaves. *American Journal of Botany* **60**, 17-33.

Hickey, L. J. (1977). Stratigraphy and paleobotany of the Golden Valley Formation (early Tertiary) of western North Dakota. *Geological Society of America Memoir* **150**.

Hickey, L. J. (1979). A revised classification of the architecture of dicotyledonous leaves. In *Anatomy of the Dicotyledons*, 2nd edition, ed. C.R Metcalfe and L. Chalk, pp. 25-39. Oxford: Clarendon.

Hickey, L. J. (1980). Paleocene stratigraphy and flora of the Clark's Fork Basin. In *Early Cenozoic paleontology and stratigraphy of the Bighorn Basin, Wyoming*, ed. P. D. Gingerich. *University of Michigan Papers in Paleontology* **24**, 33–49.

Hickey, L. J. (1981). Land plant evidence compatible with gradual, not catastrophic change at the end of the Cretaceous. *Nature* **292**, 529–531.

Hickey, L. J. (1984). Changes in the angiosperm flora across the Cretaceous–Tertiary boundary. In *Catastrophes in Earth History: The New Uniformitarianism*, ed. W. A. Berggren and J. A. Van Couvering, pp. 279–313. Princeton, NJ: Princeton University Press.

Hickey, L. J. and Doyle, J. A. (1977). Early Cretaceous fossil evidence for angiosperm evolution. *The Botanical Review* **43**, 3–104.

Hickey, L. J. and Wolfe, J. A. (1975). The bases of angiosperm phylogeny: vegetative morphology. *Annals of the Missouri Botanical Garden* **62**, 538–89.

Hickey, L. J., West, R. M., Dawson, M. R., and Choi, D. K. (1983). Arctic terrestrial biota: paleomagnetic evidence for age disparity with mid-northern latitudes during the Late Cretaceous and early Tertiary. *Science* **222**, 1153–6.

Hicks, J. F., Johnson, K. R., Obradovich, J. D., Tauxe, L., and Clark, D. (2002). Magnetostratigraphy and geochronology of the Hell Creek and basal Fort Union Formations of southwestern North Dakota and a recalibration of the age of the Cretaceous–Tertiary boundary. In *The Hell Creek Formation and the Cretaceous–Tertiary Boundary in the Northern Great Plains: An Integrated Continental Record of the End of the Cretaceous*, ed. J. H. Hartman, K. R. Johnson, and D. J. Nichols. *Geological Society of America Special Paper* **361**, 35–55.

Hicks, J. F., Johnson, K. R., Obradovich, J. D., Miggins, D. P., and Tauxe, L. (2003). Magnetostratigraphy of Upper Cretaceous (Maastrichtian) to lower Eocene strata of the Denver Basin, Colorado. *Rocky Mountain Geology* **38**, 1–27.

Hildebrand, A. R. and Boynton, W. V. (1988). Provenance of the K/T boundary layers. *Lunar and Planetary Institute Contribution* **673**, 78–9.

Hildebrand, A. R., Penfield, G. T., Kring, D. A., *et al.* (1991). Chicxulub crater: a possible Cretaceous/Tertiary boundary impact crater on the Yucatán Peninsula, Mexico. *Geology* **19**, 867–71.

Hoganson, J. W. and Murphy, E. C. (2002). Marine Breien Member (Maastrichtian) of the Hell Creek Formation in North Dakota: stratigraphy, vertebrate fossil record, and age. In *The Hell Creek Formation and the Cretaceous–Tertiary Boundary in the Northern Great Plains: An Integrated Continental Record of the End of the Cretaceous*, ed. J. H. Hartman, K. R. Johnson, and D. J. Nichols. *Geological Society of America Special Paper* **361**, 247–69.

Hollis, C. J. (2003). The Cretaceous/Tertiary boundary event in New Zealand: profiling mass extinction. *New Zealand Journal of Geology and Geophysics* **46**, 307–21.

Hollis, C. J. and Strong, C. P. (2003). Biostratigraphic review of the Cretaceous/Tertiary boundary transition, mid-Waipara River section, North Canterbury, New Zealand. *New Zealand Journal of Geology and Geophysics* **46**, 243–53.

Horrell, M. A. (1991). Phytogeography and paleoclimatic interpretation of the Maestrichtian. *Palaeogeography, Palaeoclimatology, Palaeoecology* **86**, 87–138.

Hotton, C. (2002). Palynology of the Cretaceous–Tertiary boundary in central Montana: evidence for extraterrestrial impact as a cause of the terminal Cretaceous extinctions. In *The Hell Creek Formation and the Cretaceous–Tertiary Boundary in the Northern Great Plains: An Integrated Continental Record of the End of the Cretaceous*, ed. J. H. Hartman, K. R. Johnson, and D. J. Nichols. *Geological Society of America Special Paper* **361**, 473–501.

Hotton, C. L., Leffingwell, H. A., and Skvarla, J. J. (1994). Pollen ultrastructure of Pandanaceae and the fossil genus *Pandaniidites*. In *Ultrastructure of Fossil Spores and Pollen*, ed. M. H. Kurmann and J. A. Doyle, pp. 173–91. Kew, Royal Botanic Gardens: Kew Publishing.

Hough, R. M., Wright, I. P., Pillinger, C. T., and Gilmour, I. (1995). Microdiamonds from the iridium-rich layer of the Arroyo EL Mimbral K–T boundary outcrop, N. E. Mexico. *Abstracts of the Lunar and Planetary Science Conference* **26**, 629.

Hsü, K. J. (1986). *The Great Dying*. Orlando, FL: Harcourt Brace Jovanovich.

Huber, B. T., Hodell, D. A., and Hamilton, C. P. (1995) Middle–Late Cretaceous climate of the southern high latitudes: stable isotopic evidences for minimum equator-to-pole thermal gradients. *Geological Society of America Bulletin* **107**, 1164–91.

Hunter, J. P. (1999). The radiation of Paleocene mammals with the demise of the dinosaurs: evidence from southwestern North Dakota. *Proceedings of the North Dakota Academy of Sciences* **53**, 141–4.

Hunter, J. P., Hartman, J. H., and Krause, D. W. (1997). Mammals and mollusks across the Cretaceous–Tertiary boundary from Makoshika State Park and vicinity (Williston Basin), Montana. In *Paleontology and Geology in the Northern Great Plains; a Marshall Lambert Festschrift*, ed. J. H. Hartman, pp. 61–114. Laramie, WY: University of Wyoming.

Iglesias, A., Wilf, P., Johnson, K. R., Zamuner, A. B., and Cúneo, N. R. (2006). High diversity Paleocene macrofloras from the Salamanca Formation, central Patagonia, Argentina. Abstract volume, Climate and Biota of the Early Paleogene Specialty Conference, p. 71. Bilbao, Spain, June 2006.

Iglesias, A., Wilf, P., Johnson, K. R., *et al.* (2007). A Paleocene lowland macroflora from Patagonia reveals significantly greater richness than North American analogs. *Geology* **35**, 947–50.

Izett, G. A. (1990). The Cretaceous/Tertiary boundary interval, Raton Basin, Colorado and New Mexico, and its content of shock-metamorphosed minerals; evidence relevant to the K/T boundary impact–extinction theory. *Geological Society of America Special Paper* **249**.

Jansonius, J. and McGregor, D. C., eds. (1996). *Palynology: Principles and Applications*. Dallas, TX: American Association of Stratigraphic Palynologists Foundation.

Jaramillo, C. and De la Parra, F. (2006). Palynological changes across the Cretaceous–Paleocene boundary in the Neotropics. *Geological Society of America Abstracts with Programs* **38**, 380.

Jaramillo, C., Rueda, M. J., and Mora, G. (2006). Cenozoic plant diversity in the Neotropics. *Science* **311**, 1893–6.

Jerzykiewicz, T. and Sweet, A. R. (1986). The Cretaceous–Tertiary boundary in the central Alberta Foothills. I: Stratigraphy. *Canadian Journal of Earth Sciences* **23**, 1356–74.

Johnson, K. R. (1992). Leaf-fossil evidence for extensive floral extinction at the Cretaceous–Tertiary boundary, North Dakota, USA. *Cretaceous Research* **13**, 91–117.

Johnson, K. R. (1993). Time resolution and the study of Late Cretaceous and early Tertiary megafloras. In *Taphonomic Approaches to Time Resolution in Fossil Assemblages, Short Course in Paleontology 6*, ed. S. M. Kidwell and A. K. Behrensmeyer, pp. 210–27. Knoxville, TN: The Paleontological Society.

Johnson, K. R. (2002). Megaflora of the Hell Creek and lower Fort Union formations in the western Dakotas: vegetational response to climate change, the Cretaceous–Tertiary boundary event, and rapid marine transgression. In *The Hell Creek Formation and the Cretaceous–Tertiary Boundary in the Northern Great Plains: An Integrated Continental Record of the End of the Cretaceous*, ed. J. H. Hartman, K. R. Johnson, and D. J. Nichols. *Geological Society of America Special Paper* **361**, 329–91.

Johnson, K. R. and Ellis, B. (2002). A tropical rainforest on Colorado 1.4 million years after the Cretaceous–Tertiary boundary. *Science* **296**, 2379–83.

Johnson, K. R. and Hickey, L. J. (1990). Megafloral change across the Cretaceous/Tertiary boundary in the northern Great Plains and Rocky Mountains, USA. In *Global Catastrophes in Earth History; An Interdisciplinary Conference on Impacts, Volcanism, and Mass Mortality*, ed. V. L. Sharpton and P. D. Ward. *Geological Society of America Special Paper* **247**, 433–44.

Johnson, K. R., Nichols, D. J., Attrep, Moses, Jr., and Orth, C. J. (1989). High-resolution leaf-fossil record spanning the Cretaceous/Tertiary boundary. *Nature (London)* **340**, 708–11.

Johnson, K. R., Nichols, D. J., and Hartman, J. H. (2002). Hell Creek Formation: a 2001 synthesis. In *The Hell Creek Formation and the Cretaceous–Tertiary Boundary in the Northern Great Plains: An Integrated Continental Record of the End of the Cretaceous*, ed. J. H. Hartman, K. R. Johnson, and D. J. Nichols. *Geological Society of America Special Paper* **361**, 503–10.

Johnson, K. R., Reynolds, M. L., Werth, K. W., and Thomasson, J. R. (2003). Overview of the Late Cretaceous, early Paleocene, and early Eocene megafloras of the Denver Basin, Colorado. *Rocky Mountain Geology* **38**, 101–20.

Jones, T. P. and Rowe, N. P. (1999). *Fossil Plants and Spores: Modern Techniques*. London: The Geological Society.

Kaiho, K. and Saito, T. (1986). Terminal Cretaceous sedimentary sequence recognized in the northernmost Japan based on planktonic foraminiferal evidence. *Proceedings of the Japan Academy* **62**, Ser. B, 145–8.

Kamo, S. L. and Krogh, T. E. (1995). Chicxulub crater source for shocked zircon crystals from the Cretaceous–Tertiary boundary layer, Saskatchewan; evidence from new U–Pb data. *Geology* **23**, 281–4.

Kaska, H. V. (1989). A spore and pollen zonation of Early Cretaceous to Tertiary nonmarine sediments of central Sudan. *Palynology* **13**, 79–90.

Kauffman, E. G. and Harries, P. J. (1996). The importance of crisis progenitors in recovery from mass extinctions. In *Biotic Recovery from Mass Extinction Events*, ed. M. B. Hart. *Geological Society Special Publication* **102**, 15–39.

Kauffman, E. G., Upchurch, G. R., Jr., and Nichols, D. J. (1990). The Cretaceous–Tertiary boundary interval at South Table Mountain, near Golden, Colorado. In *Extinction Events in Earth History*, ed. E. G. Kauffman and O. H. Walliser. *Lectures Notes in Earth Sciences* **30**, 365–92.

Kennedy, E. M., Spicer, R. A., and Rees, P. M. (2002). Quantitative paleoclimate estimates from Late Cretaceous and Paleocene leaf floras in the northwest of the South Island, New Zealand. *Palaeogeography, Palaeoclimatology, Palaeoecology* **184**, 321–45.

Kershaw, A. P. and Strickland, K. M. (1990). A 10 year pollen trapping record from rainforest in northeastern Queensland, Australia. *Review of Palaeobotany and Palynology* **64**, 281–8.

Knobloch, E., Kvaček, Z., Bůžek, C., Mai, D. H., and Batten, D. J. (1993). Evolutionary significance of floristic changes in the Northern Hemisphere during the Late Cretaceous and Palaeogene, with particular reference to central Europe. *Review of Palaeobotany and Palynology* **78**, 41–54.

Knowlton, F. H. (1922). The Laramie flora of the Denver Basin with a review of the Laramie problem. *US Geological Survey Professional Paper* **130**.

Knowlton, F. H. (1930). The flora of the Denver and associated formations of Colorado. *US Geological Survey Professional Paper* **155**.

Koch, B. E. (1964). Review of fossil floras and non-marine deposits of west Greenland. *Geological Society of America Bulletin* **75**, 535–48.

Kodrul, T. (2004). The middle Tsagayan flora of Amur River region. In *Proceedings of the 3rd Symposium on Cretaceous Biota and the K/T Boundary in Heilongjiang River Area*, ed. G. Sun, Y. W. Sun, M. Akhmetiev, and A. R. Ashraf, pp. 17–22. Changchun, China: Jilin University Research Center of Palaeontology and Stratigraphy.

Koeberl, C. and MacLeod, K. G. (2002). *Catastrophic Events and Mass Extinctions: Impacts and Beyond*. Geological Society of America Special Paper 356.

Krassilov, V. A. (1976). *Tsagajan flora of Amur Province*. Moscow: Nauka. [in Russian]

Krassilov, V. A. (1978). Late Cretaceous gymnosperms from Sakhalin and the terminal Cretaceous event. *Palaeontology* **21**, 893–905.

Krassilov, V. A. (1979) *The Cretaceous flora of Sakhalin*. Moscow: Nauka. [in Russian]

Krassilov, V. A. (1987). Palaeobotany of the Mesophyticum: state of the art. *Review of Palaeobotany and Palynology* **50**, 231–54.

Krassilov, V. A. (2003). *Terrestrial Paleoecology and Global Change*. Moscow: Pensoft Publishers.

Kroeger, T. J. (2002). Palynology of the Hell Creek Formation (Upper Cretaceous, Maastrichtian) in northwestern South Dakota: effects of paleoenvironment on the composition of palynomorph assemblages. In *The Hell Creek Formation and the Cretaceous–Tertiary Boundary in the Northern Great Plains: An Integrated Continental Record of the End of the Cretaceous*, ed. J. H. Hartman, K. R. Johnson, and D. J. Nichols. *Geological Society of America Special Paper* **361**, 457–72.

Labandeira, C. C., Johnson, K. R., and Wilf, P. (2002). Impact of the terminal Cretaceous event on plant–insect associations. *Proceedings of the National Academy of Science* **99**, 2061–6.

Lee, W. T. and Knowlton, F. H. (1917). Geology and paleontology of Raton Mesa and other regions in Colorado and New Mexico. *US Geological Survey Professional Paper* **101**.

Leffingwell, H. A. (1970). Palynology of the Lance (Late Cretaceous) and Fort Union (Paleocene) formations of the type Lance area, Wyoming. In *Symposium on Palynology of the Late Cretaceous and Early Tertiary*, ed. R. M. Kosanke and A. T. Cross. *Geological Society of America Special Paper* **127**, 1–64.

Lerbekmo, J. F. (1985). Magnetostratigraphic and biostratigraphic correlations of Maastrichtian to early Paleocene strata between south–central Alberta and southwestern Saskatchewan. *Bulletin of Canadian Petroleum Geology* **33**, 213–26.

Lerbekmo, J. F. and Coulter, K. C. (1984). Magnetostratigraphic and biostratigraphic correlations of Late Cretaceous to early Paleocene strata between Alberta and North Dakota. In *The Mesozoic of Middle North America*, ed. D. F. Stott and D. J. Glass. *Canadian Society of Petroleum Geologists Memoir* **9**, 313–17.

Lerbekmo, J. F., Evans, M. E., and Baadsgaard, H. (1979). Magnetostratigraphy, biostratigraphy and geochronology of Cretaceous–Tertiary boundary sediments, Red Deer Valley. *Nature (London)* **279**, 26–30.

Lerbekmo, J. F., Sweet, A. R., and St. Louis, R. M. (1987). The relationship between the iridium anomaly and palynological floral events at three Cretaceous–Tertiary boundary localities in western Canada. *Geological Society of America Bulletin* **99**, 325–30.

Lerbekmo, J. F., Sweet, A. R., and Braman, D. R. (1995). Magnetostratigraphy of late Maastrichtian to early Paleocene strata of the Hand Hills, south central Alberta, Canada. *Bulletin of Canadian Petroleum Geology* **43**, 35–43.

Lerbekmo, J. F., Sweet, A. R., and Duke, M. J. M. (1996). A normal polarity subchron that embraces the K/T boundary: a measure of sedimentary continuity across the boundary and synchroneity of boundary events. In *The Cretaceous–Tertiary Event and Other Catastrophes in Earth History*, ed. G. Ryder, D. Fastovsky, and S. Gartner. *Geological Society of America Special Paper* **370**, 465–86.

Lidgard, S. and Crane, P. R. (1990). Angiosperm diversification and Cretaceous floristic trends: a comparison of palynofloras and leaf macrofloras. *Paleobiology* **16**, 77–93.

Liu, M. (1983). The late Upper Cretaceous to Palaeocene spore pollen assemblages from the Furao area, Heilongjiang Province. *Bulletin of the Shenyang Institute of Geology and Mineral Resources, Chinese Academy of Geological Sciences* **7**, 99–131. [in Chinese with English abstract]

López-Martinez, N., Ardevol, L., Arribas, M. E., Civis, J., and Gonzalez-Delgado, A. (1998). The geological record in non-marine environments around the K/T boundary (Tremp Formation, Spain). *Bulletin de la Société Géologique de France* **169**, 11–20.

López-Martinez, N., Fernández-Marrón, M. T., and Valle, M. F. (1999). The succession of vertebrates and plants across the Cretaceous–Tertiary boundary in the Tremp Formation, Ager Valley (south-central Pyrenees, Spain). *Geobios* **32**, 617–27.

Maley, J. (1996). The African rain forest – main characteristics of changes in vegetation and climate from the Upper Cretaceous to the Quaternary. *Proceedings of the Royal Society of Edinburgh* **104B**, 31–73.

Manchester, S. R., Crane, P. R., and Golovneva, L. B. (1999). An extinct genus with affinities to extant *Davidia* and *Camptotheca* (Cornales) from the Paleocene of North America and eastern Asia. *International Journal of Plant Sciences* **160**, 188–207.

Mangin, J.-P. (1957). Remarques sur le terme Paléocène et sur la limite Crétacé–Tertiare. *Comptes rendus des séances de la Société Géologique de France* **14**, 319–21.

Markevich [Markevitch], V. S. (1994). Palynological zonation of the continental Cretaceous and lower Tertiary of eastern Russia. *Cretaceous Research* **15**, 165–77.

Markevich, V. S. and Bugdaeva, E. V. (2001). The Maastrichtian flora and dinosaurs of the Russian Far East. In *Proceedings of the IXth International Palynological Congress*, ed. D. K. Goodman and R. T. Clarke, pp. 139–48. Dallas, TX: American Association of Stratigraphic Palynologists Foundation.

Markevich, V. S., Bolotsky, Y. L., and Bugdaeva, E. V. (1994). The Kundur dinosaur-bearing locality in the Priamurye (Amur River region). *Tikhookeanskaya Geologiya (Pacific Geology)* **6**, 96–107. [in Russian]

Markevich [Markevitch], V. S., Bugdaeva, E. V., and Bolotsky, Y. L. (2000). Palynological evidence of vegetational change and dinosaur extinction in the Amur region. *Paleontological Journal* **34**, Suppl. 1, S50–3.

Markevich, V. S., Bugdaeva, E. V., and Ashraf, A. R. (2004). Results of study of Arkhara-Boguchan coal field. In *Proceedings of the 3rd Symposium on Cretaceous Biota and K/T Boundary in Heilongjiang River Area*, ed. G. Sun, Y. W. Sun, M. A. Akhmetiev, and R. A. Ashraf, pp. 41–4. Changchun, China: Jilin University Research Center of Palaeontology and Stratigraphy.

Mateer, N. and Chen, P. (1992). A review of the nonmarine Cretaceous–Tertiary transition in China. *Cretaceous Research* **13**, 81–90.

Mayr, C., Thümmler, B., Windmaier, G., *et al.* (1999). New data about the Maastrichtian/Danian transition in the Southern Pyrenees (Ager Basin, Catalonia, Spain). *Revista española de micropaleontologia* **31**, 357–68.

McIver, E. E. (1999). Paleobotanical evidence for ecosystem disruption at the Cretaceous–Tertiary boundary from Wood Mountain, Saskatchewan, Canada. *Canadian Journal of Earth Sciences* **36**, 775–89.

McIver, E. E. (2002). The paleoenvironment of *Tyrannosaurus rex* from southwestern Saskatchewan, Canada. *Canadian Journal of Earth Science* **39**, 207–21.

McIver, E. E. and Basinger, J. F. (1993). Flora of the Ravenscrag Formation (Paleocene), southwestern Saskatchewan, Canada. *Palaeontographica Canadiana* **10**, 1–64.

Médus, J., Feist, M., Rocchia, R., *et al.* (1988). Prospects for recognition of the palynological Cretaceous/Tertiary boundary and an iridium anomaly in nonmarine facies of the eastern Spanish Pyrenees: a preliminary report. *Newsletters on Stratigraphy* **18**, 123–38.

Médus, J., Colombo, F., and Durand, J. P. (1992). Pollen and spore assemblages of the uppermost Cretaceous continental formations of south-eastern France and north-eastern Spain. *Cretaceous Research* **13**, 119–32.

Meldahl, K. H. (1990). Sampling, species abundance, and the stratigraphic signature of mass extinction – a test using Holocene tidal flat mollusks. *Geology* **18**, 890–3.

Melosh, H. J., Schneider, N. M., Zahnle, K. J., and Latham, D. (1990). Ignition of global wildfires at the Cretaceous/Tertiary boundary. *Nature* **343**, 251–4.

Méon, H. (1990). Palynologic studies of the Cretaceous–Tertiary boundary interval at El Kef outcrop, northwestern Tunisia: paleogeographic implications. In *Proceedings of the 7th International Palynological Congress*, ed. E. M. Truswell and J. A. K. Owen. *Review of Palaeobotany and Palynology* **65**, 85–94.

Méon, H. (1991). Études sporopollinique à la limite Crétacé-Tertiare: la coupe de Kef (Tunisie nord-occidentale); étude systématique, stratigraphie, paléogéographie et évolution climatique. *Palaeonotographica, Abteilung B* **223**, 107–68.

Mtchedlishvili, N. D. (1964). The significance of angiospermous plants for stratigraphy of Upper Cretaceous deposits. *Trudy VNIGRI [All-Union Oil and Geological Survey Research Institute]* **239**, 5–37. [in Russian; translated by Canadian Multilingual Services Division in Geological Survey of Canada, 1976]

Muller, J. (1970). Palynological evidence on early differentiation of angiosperms. *Biological Review* **45**, 417–50.

Muller, J., De Di Giacomo, E., and Van Erve, A. W. (1987). A palynological zonation for the Cretaceous, Tertiary, and Quaternary of northern South America. *American Association of Stratigraphic Palynologists Contributions Series* **19**, 7–71.

Murphy, E. C., Nichols, D. J., Hoganson, J. W., and Forsman, N. F. (1995). *The Cretaceous–Tertiary Boundary in South-Central North Dakota*. North Dakota Geological Survey Report of Investigation 98.

Nandi, B. (1990). Palynostratigraphy of Upper Cretaceous sediments, Meghalaya, northeastern India. *Review of Palaeobotany and Palynology* **65**, 119–29.

Nandi, B. and Chattopadhyay, S. (2002). Triprojectaperturate pollen grains from the Late Cretaceous sediments of Meghalaya. *Acta Palaeontologica Sinica* **41**, 601–10.

Newman, K. R. (1979). Cretaceous/Paleocene boundary in the Denver Formation at Golden, Colorado, USA. In *Cretaceous–Tertiary Boundary Events, Symposium II, Proceedings*, ed. W. K. Christensen and T. Birkelund. *University of Copenhagen Geological Museum, Contributions to Palaeontology* **294**, 246–8.

Nichols, D. J. (1990). Geologic and biostratigraphic framework of the non-marine Cretaceous–Tertiary boundary interval in western North America. In *Proceedings of the 7th International Palynological Congress*, ed. E. M. Truswell and J. A. K. Owen. *Review of Palaeobotany and Palynology* **65**, 75–84.

Nichols, D. J. (1994). A revised palynostratigraphic zonation of the nonmarine Upper Cretaceous, Rocky Mountain region, United States. In *Mesozoic Systems of the Rocky Mountain Region, USA*, pp. 503–521. Denver, CO: Rocky Mountain Section of the Society for Sedimentary Geology (SEPM).

Nichols, D. J. (2002a). Palynology and microstratigraphy of the Hell Creek Formation in North Dakota: a microfossil record of plants at the end of Cretaceous time. In *The Hell Creek Formation and the Cretaceous–Tertiary Boundary in the Northern Great Plains: An Integrated Continental Record of the End of the Cretaceous*, ed. J. H. Hartman, K. R. Johnson, and D. J. Nichols. *Geological Society of America Special Paper* **361**, 393–456.

Nichols, D. J. (2002b). Correlation of palynostratigraphy and magnetostratigraphy in Maastrichtian-Eocene strata in the Denver Basin, Colorado. *Geological Society of America Abstracts with Programs* **34**, 508.

Nichols, D. J. (2003). Palynostratigraphic framework for age determination and correlation of the nonmarine lower Cenozoic of the Rocky Mountains and Great Plains region. In *Cenozoic Systems of the Rocky Mountain Region*, ed. R. G. Raynolds and R. M. Flores, pp. 107-34. Denver, CO: Rocky Mountain Section of the Society for Sedimentary Geology (SEPM).

Nichols, D. J. and Fleming, R. F. (1990). Plant microfossil record of the terminal Cretaceous event in the western United States and Canada. In *Global Catastrophes in Earth History; An Interdisciplinary Conference on Impacts, Volcanism, and Mass Mortality*, ed. V. L. Sharpton and P. D. Ward. *Geological Society of America Special Paper* **247**, 445-55.

Nichols, D. J. and Fleming, R. F. (2002). Palynology and palynostratigraphy of Maastrichtian, Paleocene, and Eocene strata in the Denver Basin, Colorado. *Rocky Mountain Geology* **37**, 135-63.

Nichols, D. J. and Johnson, K. R. (2002). Palynology and microstratigraphy of Cretaceous-Tertiary boundary sections in southwestern North Dakota. In *The Hell Creek Formation and the Cretaceous-Tertiary Boundary in the Northern Great Plains: An Integrated Continental Record of the End of the Cretaceous*, ed. J. H. Hartman, K. R. Johnson, and D. J. Nichols. *Geological Society of America Special Paper* **361**, 95-143.

Nichols, D. J. and Pillmore, C. L. (2000). Palynology of the K-T boundary in the Raton Basin, Colorado and New Mexico - new data and interpretations from the birthplace of K-T plant microfossil studies in nonmarine rocks. *Conference on Catastrophic Events and Mass Extinctions: Impacts and Beyond, Vienna, Austria*. See www.lpi.usra.edu/meetings/impact2000/pdf/3120.pdf

Nichols, D. J. and Sweet, A. R. (1993). Biostratigraphy of Upper Cretaceous non-marine palynofloras in a north-south transect of the Western Interior basin. In *Evolution of the Western Interior Basin*, ed. W. G. E. Caldwell and E. G. Kauffman. *Geological Society of Canada Special Paper* **39**, 539-84.

Nichols, D. J., Jarzen, D. M., Orth, C. J., and Oliver, P. Q. (1986). Palynological and iridium anomalies at Cretaceous-Tertiary boundary, south-central Saskatchewan. *Science* **231**, 714-17.

Nichols, D. J., Wolfe, J. A., and Pocknall, D. T. (1988). Latest Cretaceous and early Tertiary history of vegetation in the Powder River basin, Montana and Wyoming. In *Geological Society of America 1888-1988 Centennial Meeting, Denver, Colorado, Field Trip Guidebook 1988*, ed. G. S. Holden. *Colorado School of Mines Professional Contributions* **12**, 205-10, 222-6.

Nichols, D. J., Fleming, R. F., and Frederiksen, N. O. (1990). Palynological evidence of effects of the terminal Cretaceous event in western North America. In *Extinction Events in Earth History*, ed. E. G. Kauffman and O. H. Walliser. *Lecture Notes in Earth Sciences* **30**, 351-64.

Nichols, D. J., Brown, J. L., Attrep, M., Jr., and Orth, C. J. (1992a). A new Cretaceous-Tertiary boundary locality in the western Powder River Basin, Wyoming: biological and geological implications. *Cretaceous Research* **13**, No. 1, 3-30.

Nichols, D. J., Hickey, L. J., McWeeney, L. J., and Wolfe, J. A. (1992b). Plants at the K/T boundary; discussion and reply. *Nature (London)* **356**, 295–6.

Obradovich, J. D. (1993). A Cretaceous time scale. In *Evolution of the Western Interior Basin*, ed. W. G. E. Caldwell and E. G. Kauffman. *Geological Society of Canada Special Paper* **39**, 379–96.

Obradovich, J. D. (2002). Geochronology of Laramide synorogenic strata in the Denver Basin, Colorado. *Rocky Mountain Geology* **37**, 165–71.

Odin, G. S. (1996). Definition of a Global Boundary Stratotype Section and Point for the Campanian/Maastrichtian boundary. *Bulletin de l'Institut Royal des Sciences Naturelles de Belgique, Sciences de la Terre* **66 Supp.**, 111–17.

Officer, C. and Page, J. (1996). *The Great Dinosaur Extinction Controversy*. Reading, MA: Addison-Wesley.

Orth, C. J., Gilmore, J. S., Knight, J. D., *et al.* (1981). An iridium abundance anomaly at the palynological Cretaceous–Tertiary boundary in northern New Mexico. *Science* **214**, 1341–3.

Orth, C. J., Gilmore, J. S., Knight, J. D., *et al.* (1982). Iridium abundance measurements across the Cretaceous/Tertiary boundary in the San Juan and Raton Basins of northern New Mexico. In *Geological Implications of Impacts of Large Asteroids and Comets on the Earth*, ed. L. T. Silver and P. H. Schultz. *Geological Society of America Special Paper* **190**, 423–33.

Orth, C. J., Gilmore, J. S., and Knight, J. D. (1987). Iridium anomaly at the Cretaceous–Tertiary boundary in the Raton Basin. *New Mexico Geological Society Guidebook, 38th Field Conference*, pp. 265–9. Socorro, NM: New Mexico Geological Society.

Pearson, D. A., Schaefer, T., Johnson, K. R., and Nichols, D. J. (2001). Palynologically calibrated vertebrate record from North Dakota consistent with abrupt dinosaur extinction at the Cretaceous–Tertiary boundary. *Geology* **29**, 39–42.

Pearson, D. A., Schaefer, T., Johnson, K. R., Nichols, D. J., and Hunter, J. P. (2002). Vertebrate biostratigraphy of the Hell Creek Formation in southwestern North Dakota and northwestern South Dakota. In *The Hell Creek Formation and the Cretaceous–Tertiary Boundary in the Northern Great Plains: An Integrated Continental Record of the End of the Cretaceous*, ed. J. H. Hartman, K. R. Johnson, and D. J. Nichols. *Geological Society of America Special Paper* **361**, 145–67.

Peppe, D. J., Evans, D. A. D., and Hickey, L. J. (2005). Magnetostratigraphy and megaflora of the lower Paleocene Fort Union Formation along the southwestern margin of the Williston Basin, North Dakota. *Geological Society of America Abstracts with Programs* **37**, 160.

Pillmore, C. L. and Fleming, F. (1990). The Cretaceous–Tertiary boundary in the Raton Basin, New Mexico and Colorado. *New Mexico Geological Survey, 41st Field Conference, Sangre de Cristo Mountains, New Mexico*, pp. 327–31. Socorro, NM: New Mexico Geological Society.

Pillmore, C. L., Tschudy, R. H., Orth, C. J. *et al.* (1984). Geologic framework of nonmarine Cretaceous–Tertiary boundary sites, Raton Basin, New Mexico and Colorado. *Science* **223**, 1180–3.

Pillmore, C. L., Flores, R. M., and Fleming, R. F. (1988). Field guide to the continental Cretaceous-Tertiary boundary in the Raton Basin, Colorado and New Mexico. In *Geological Society of America Field Trip Guidebook 1988*, ed. G. S. Holden. *Colorado School of Mines Professional Contribution* **12**, 227–58.

Pillmore, C. L., Nichols, D. J., and Fleming, R. F. (1999). Field guide to the continental Cretaceous-Tertiary boundary in the Raton Basin, Colorado and New Mexico. In *Colorado and Adjacent Areas*, ed. D. R. Lageson, A. P. Lester, and B. D. Trudgill. *Geological Society of America Field Guide* **1**, 135–55.

Pocknall, D. T., Erlich, R. N., Stein, J. A., and Lorente, M. A. (2001). The palynofloral succession across the Cretaceous to Paleocene transition zone, Merida Andes, western Venezuela. In *Proceedings of the IXth International Palynological Congress*, ed. D. K. Goodman and R. T. Clarke, pp. 171–9. Dallas, TX: American Association of Stratigraphic Palynologists Foundation.

Pole, M. (1992). Cretaceous macrofloras of eastern Otago, New Zealand: angiosperms. *Australian Journal of Botany* **40**, 169–206.

Pope, K. O. (2002). Impact dust not the cause of the Cretaceous–Tertiary mass extinction. *Geology* **30**, 99–102.

Pope, K. O., Baines, K. H., Ocampo, A. C., and Ivanov, B. A. (1997). Energy, volatile production, and climatic effects of the Chicxulub Cretaceous/Tertiary impact. *Journal of Geophysical Research* **102**, 21645–64.

Powell, J. L. (1998). *Night comes to the Cretaceous. Comets, Craters, Controversy, and the Last Days of the Dinosaurs*. New York, NY: W.H. Freeman.

Prinn, R. G. and Fegley, B., Jr. (1987). Bolide impacts, acid rain, and biospheric trauma at the Cretaceous–Tertiary boundary. *Earth and Planetary Science Letters* **83**, 1–15.

Raine, J. I. (1984). Outline of a palynological zonation of Cretaceous to Paleogene terrestrial sediments in West Coast region, South Island, New Zealand. *New Zealand Geological Survey Report* **109**, 1–82.

Raynolds, R. G. (2002). Upper Cretaceous and Tertiary stratigraphy of the Denver Basin, Colorado. *Rocky Mountain Geology* **37**, 105–9.

Raynolds, R. G. and Johnson, K. R. (2002). Drilling of the Kiowa core, Elbert County, Colorado. *Rocky Mountain Geology* **37**, 111–34.

Raynolds, R. G. and Johnson, K. R. (2003). Synopsis of the stratigraphy and paleontology of the uppermost Cretaceous and lower Tertiary strata in the Denver Basin, Colorado. *Rocky Mountain Geology* **38**, 171–81.

Richards, P. W. (1952). *The Tropical Rain Forest*. Cambridge: Cambridge University Press.

Rigby, J. K., Jr., Snee, L. W., Unruh, D. M., *et al.* (1993). $^{40}Ar/^{40}39$ and U-Pb dates for dinosaur extinction, Nanxiong basin, Guangdong Province, People's Republic of China. *Geological Society of America Abstracts with Programs* **25**, 296.

Roberts, L. N. R., and Kirschbaum, M. A. (1995). Paleogeography of the Late Cretaceous of the Western Interior of middle North America - coal distribution and sediment accumulation. *US Geological Survey Professional Paper* **1561**.

Robertson, D. D., McKenna, M. C., Toon, O. B., Hope, S., and Lillegraven, J. A. (2004). Survival in the first hours of the Cenozoic. *Geological Society of America Bulletin* **116**, 760–8.

Robin, E., Boclet, D., Bonté, Ph., *et al.* (1991). The stratigraphic distribution of Ni-rich spinels in Cretaceous–Tertiary boundary rocks at El Kef (Tunisia), Caravaca (Spain) and Hole 761 C (Leg 122). *Earth and Planetary Science Letters* **107**, 715–21

Rouse, G. E. (1957). The application of a new nomenclatural approach to Upper Cretaceous plant microfossils from western Canada. *Canadian Journal of Botany* **35**, 349–75.

Russell, D. A. and Manabe, M. (2002). Synopsis of the Hell Creek (uppermost Cretaceous) dinosaur assemblage. In *The Hell Creek Formation and the Cretaceous–Tertiary Boundary in the Northern Great Plains: An Integrated Continental Record of the End of the Cretaceous*, ed. J. H. Hartman, K. R. Johnson, and D. J. Nichols. *Geological Society of America Special Paper* **361**, 169–76.

Russell, D. A., Russell, D. F., and Sweet, A. R. (1993). The end of the dinosaurian era in Nanxiong Basin. *Vertebrata PalAsiatica* **31**, 139–45.

Ryder, G., Fastovsky, D., and Gartner, S., eds. (1996). *The Cretaceous–Tertiary Event and Other Catastrophes in Earth History*. Geological Society of America Special Paper 307.

Saito, T., Yamanoi, T., and Kaiho, K. (1986). End-Cretaceous devastation of terrestrial flora in the boreal Far East. *Nature* **323**, 253–5.

Salard-Cheboldaeff, M. (1990). Intertropical African palynostratigraphy from Cretaceous to late Quaternary times. *Journal of African Earth Sciences* **11**, 1–24.

Samoilovich, S. R., Mtchedlishvili, N. D., Rusakova, L. I., and Ishchurzhinskaia, A. B., eds. (1961). Pollen and spores of western Siberia, Jurassic to Paleocene. *Trudy VNIGRI [All-Union Oil and Geological Survey Research Institute]* **177**. [in Russian]

Schimmelmann, A. and DeNiro, M. J. (1984). Elemental and stable isotope variations of organic matter from a terrestrial sequence containing the Cretaceous/Tertiary boundary at York Canyon, New Mexico. *Earth and Planetary Science Letters* **68**, 392–8.

Sewall, J. O. and Sloan, L. C. (2006). Come a little closer; a high-resolution climate study of the early Paleogene Laramide foreland. *Geology* **34**, 81–4.

Sharpton, V. L. and Ward, P. D., eds. (1990). *Global Catastrophes in Earth History; An Interdisciplinary Conference on Impacts, Volcanism, and Mass Mortality*. Geological Society of America Special Paper 247.

Sheehan, P. M., Fastovsky, D. E., Barreto, C., and Hoffman, R. G. (2000). Dinosaur abundance was not declining in a "3 m gap" at the top of the Hell Creek Formation, Montana and North Dakota. *Geology* **28**, 523–6.

Shoemaker, E. M., Pillmore, C. L., and Peacock, E. W. (1987). Remanent magnetism of rocks of latest Cretaceous and earliest Tertiary age from drill core at York Canyon, New Mexico. In *The Cretaceous–Tertiary Boundary in the San Juan and Raton Basins, New Mexico and Colorado*, ed. J. E. Fassett and J. K. Rigby, Jr. *Geological Society of America Special Paper* **208**, 131–50.

Shoemaker, R. E. (1966). Fossil leaves of the Hell Creek and Tullock Formations of eastern Montana. *Palaeontographica, Abteilung B* **119**, 54–75.

Shukla, A. D. and Shukla, P. N. (2002). Comments on 'No K/T boundary at Anjar, Gujarat, India: evidence from magnetic susceptibility and carbon isotopes' by

H. J. Hansen, D. M. Mohabey, and P. Toft. *Proceedings of the Indian Academy of Science (Earth and Planetary Sciences)* **111**, 489-91.

Signor, P. W. and Lipps, J. H. (1982). Sampling bias, gradual extinction patterns, and catastrophes in the fossil record. In *Geological implications of impacts of large asteroids and comets on the Earth*, ed. L. T. Silver and P. H. Schultz. *Geological Society of America Special Paper* **190**, 291-6.

Sigurdsson, H., D'Hondt, S., and Carey, S. (1992). The impact of the Cretaceous/Tertiary bolide on evaporite terrane and generation of major sulfuric acid aerosol. *Earth and Planetary Science Letters* **109**, 543-59.

Silver, L. T. and Schultz, P. H., eds. (1982). *Geological Implications of Impacts of Large Asteroids and Comets on the Earth*. Geological Society of America Special Paper 190.

Singh, R. S., Kar, R., and Prasad, G. V. R. (2006). Palynological constraints on the age of mammal-yielding Deccan intertrappean beds of Naskal, Rangareddi district, Andhra Pradesh. *Current Science* **90**, 1281-5.

Smit, J. (1999). The global stratigraphy of the Cretaceous-Tertiary boundary impact ejecta. *Annual Review of Earth and Planetary Sciences* **27**, 75-113.

Smit, J. and Van der Kaars, S. (1984). Terminal Cretaceous extinctions in the Hell Creek area, Montana: compatible with catastrophic extinction. *Science* **223**, 1177-9.

Smit, J., Van der Kaars, W. A., and Rigby, J. K., Jr. (1987). Stratigraphic aspects of the Cretaceous-Tertiary boundary in the Bug Creek area of eastern Montana, USA. *Memoirs de la Société Géologique de France N.S.* **150**, 53-73.

Song, Z. and Huang, F. (1997). Comparison of palynomorph assemblages from the Cretaceous/Tertiary boundary interval in western Europe, northwest Africa and southeast China. *Cretaceous Research* **18**, 865-72.

Song, Z., Zheng, Y., Liu, J., et al. (1980). Cretaceous-Tertiary sporo pollen assemblages of northern Jiangsu. Abstract volume, 5th International Palynological Conference. Cambridge, UK, June 1980.

Song Z., Li, W., and He, C. (1983). Cretaceous and Palaeogene palynofloras and distribution of organic rocks in China. *Scientia Sinica B* **26**, 538-49.

Song, Z., Zheng, Y., and Liu, J. (1995). Palynological assemblages across the Cretaceous/Tertiary boundary in northern Jiangsu, eastern China. *Cretaceous Research* **16**, 465-82.

Spicer, R. A., Davies, K. S., and Herman, A. B. (1994). Circum-Arctic plant fossils and the Cretaceous-Tertiary transition. In *Cenozoic Plants and Climates of the Arctic*, ed. M. C. Boulter, and H. C. Fisher, pp. 161-74. Heidelberg: Springer-Verlag.

Stanley, E. A. (1965). Upper Cretaceous and Paleocene plant microfossils and Paleocene dinoflagellates and hystrichosphaerids from northwestern South Dakota. *Bulletins of American Paleontology* **49**, 177-384.

Stets, J., Ashraf, A. R., Erben, H. K., Hahn, G. et al. (1996). The Cretaceous-Tertiary boundary in the Nanxiong Basin (continental facies, southeast China). In *Cretaceous-Tertiary Mass extinctions: Biotic and Environmental Changes*, ed. N. McLeod and G. Keller, pp. 349-71. New York, NY: W.W. Norton.

Sun, G., Akhmetiev, M., Dong, Z. M., et al. (2002). In search of the Cretaceous-Tertiary boundary in Heilongjiang River area of China. *Journal of Geoscience Research in Northeast Asia* **5**, 105-13.

Sun, G., Akhmetiev, M., Ashraf, A. R., *et al.* (2004). Recent advance on the research of Cretaceous-Tertiary boundary in Jiayin of Heilongjiang, China. In *Proceedings of the 3rd Symposium on Cretaceous Biota and the K/T Boundary in Heilongjiang River Area*, ed. G. Sun, Y. W. Sun, M. Akhmetiev, and A. R. Ashraf, pp. 1–6. Changchun, China: Jilin University Research Center of Palaeontology and Stratigraphy.

Sweet, A. R. and Braman, D. R. (1992). The K–T boundary and contiguous strata in western Canada: interactions between paleoenvironments and palynological assemblages. *Cretaceous Research* **13**, 31–79.

Sweet, A. R. and Braman, D. R. (2001). Cretaceous-Tertiary palynofloral perturbations and extinctions within the *Aquilapollenites* phytogeographic province. *Canadian Journal of Earth Sciences* **38**, 249–69.

Sweet, A. R. and Jerzykiewicz, T. (1987). Sedimentary facies and environmentally controlled palynological assemblages: their relevance to floral changes at the Cretaceous-Tertiary boundary. In *Fourth Symposium on Mesozoic Terrestrial Ecosystems*, ed. P. J. Currie and E. H. Koster. *Occasional Paper of the Tyrrell Museum of Palaeontology* **3**, 206–11.

Sweet, A. R., Braman, D. R., and Lerbekmo, J. F. (1990). Palynofloral response to K/T boundary events; a transitory interruption within a dynamic system. In *Global Catastrophes in Earth History; An Interdisciplinary Conference on Impacts, Volcanism, and Mass Mortality*, ed. V. L. Sharpton and P. D. Ward. *Geological Society of America Special Paper* **247**, 457–69.

Sweet, A. R., Braman, D. R., and Lerbekmo, J. F. (1999). Sequential palynological changes across the composite Cretaceous-Tertiary boundary claystone and contiguous strata, western Canada and Montana, U.S.A. *Canadian Journal of Earth Sciences* **36**, 743–68.

Swisher, C. C., Dingus, L., and Butler, R. F. (1993). $^{40}Ar/^{39}Ar$ dating and magnetostratigraphic correlation of the terrestrial Cretaceous–Paleogene boundary and Puercan Mammal Age, Hell Creek–Tullock formations, eastern Montana. *Canadian Journal of Earth Sciences* **30**, 1981–95.

Takahashi, K. and Shimono, S. (1982). Maastrichtian microflora of the Miyadani-gawa Formation in the Hida District, central Japan. *Bulletin of the Faculty of Liberal Arts, Nagasaki University, Natural Science* **22**, 11–118.

Takahashi, K. and Yamanoi, T. (1992). Palynologic study of Kawaruppu K/T boundary samples in eastern Hokkaido. *Bulletin of the Faculty of Liberal Arts, Nagasaki University, Natural Science* **32**, 187–220. [in Japanese with English abstract]

Takhtajan, A. L. (1980). Outline of the classification of flowering plants (Magnoliophyta). *Botanical Review* **46**, 1–225.

Taylor, L. H., Jacobs, L. L., and Downs, W. R. (2006). A review of the Cretaceous-Paleogene boundary in the Nanxiong Basin. In *Papers from the 2005 Heyuan International Dinosaur Symposium*, ed. J. C. Lü, Y. Kobayashi, D. Huang, and Y.-N. Lee, pp. 39–59. Beijing: Geological Publishing House.

Toon, O. B., Zahnle, K., Turco, R. P., and Covey, C. (1994). Environmental perturbations caused by impacts. In *Hazards Due to Comets and Asteroids*, ed. T. Gehrels, pp. 791–826. Tucson, AZ: University of Arizona Press.

Traverse, A. (1988a). Plant evolution dances to a different beat; plant and animal evolutionary mechanisms compared. *Historical Biology* **1**, 277–301.

Traverse, A. (1988b). *Paleopalynology*. Boston, MA: Unwin Hyman.

Tschudy, B. D. and Leopold, E. B. (1970). *Aquilapollenites* (Rouse) Funkhouser – selected Rocky Mountain taxa and their stratigraphic ranges. In *Symposium on Palynology of the Late Cretaceous and Early Tertiary*, ed. R. M. Kosanke and A. T. Cross. *Geological Society of America Special Paper* **127**, 113–67.

Tschudy, R. H. (1970). Palynology of the Cretaceous–Tertiary boundary in the northern Rocky Mountain and Mississippi Embayment regions. In *Symposium on Palynology of the Late Cretaceous and Early Tertiary*, ed. R. M. Kosanke and A. T. Cross. *Geological Society of America Special Paper* **127**, 65–111.

Tschudy, R. H. (1984). Palynological evidence for change in continental floras at the Cretaceous Tertiary boundary. In *Catastrophes in Earth History: The New Uniformitarianism*, ed. W. A. Berggren and J. A. Van Couvering, pp. 315–37. Princeton, NJ: Princeton University Press.

Tschudy, R. H. and Tschudy, B. D. (1986). Extinction and survival of plant life following the Cretaceous–Tertiary boundary event, Western Interior, North America. *Geology* **14**, 667–70.

Tschudy, R. H., Pillmore, C. L., Orth, C. J., Gilmore, J. S., and Knight, J. D. (1984). Disruption of the terrestrial plant ecosystem at the Cretaceous–Tertiary boundary, Western Interior. *Science* **225**, 1030–2.

Upchurch, G. R., Jr. and Wolfe, J. A. (1993). Cretaceous vegetation of the Western Interior and adjacent regions of North America. In *Evolution of the Western Interior Basin*, ed. W. G. E. Caldwell and E. G. Kauffman. *Geological Society of Canada Special Paper* **39**, 243–81.

Vajda, V. and McLoughlin, S. (2004). Fungal proliferation at the Cretaceous–Tertiary boundary. *Science* **303**, 1489.

Vajda, V. and McLoughlin, S. (2007). Extinction and recovery patterns of the vegetation across the Cretaceous–Palaeogene boundary – a tool for unravelling the causes of the end-Permian mass-extinction. *Review of Palaeobotany and Palynology* **144**, 99–112.

Vajda, V. and Raine, J. I. (2003). Pollen and spores in marine Cretaceous/Tertiary boundary sediments at mid-Waipara River, North Canterbury, New Zealand. *New Zealand Journal of Geology and Geophysics* **46**, 255–73.

Vajda, V., Raine, J. I., and Hollis, C. J. (2001). Indication of global deforestation at the Cretaceous–Tertiary boundary by New Zealand fern spike. *Science* **294**, 1700–2.

Vajda, V., Raine, J. I., Hollis, C. J., and Strong, C. P. (2004). Global effects of the Chicxulub impact on terrestrial vegetation – review of the palynological record from New Zealand Cretaceous/Tertiary boundary. In *Cratering in marine environments and on ice*, ed. H. Dypvik, M. Burchell, and P. Claeys, pp. 57–74. Berlin: Springer-Verlag.

Van Itterbeeck, J., Bolotsky, Y., Bultynck, P., and Godefroit, P. (2005). Stratigraphy, sedimentology and palaeoecology of the dinosaur-bearing Kundur section (Zeya-Bureya Basin, Amur Region, Far Eastern Russia). *Geological Magazine* **6**, 1–16.

Venkatesan, T. R., Pande, K., and Gopalan, K. (1993). Did Deccan volcanism predate the K/T transition? *Earth and Planetary Science Letters* **119**, 181–9.

Venkatesan, T. R., Pande, K., and Ghevariya, Z. G. (1996). ^{40}Ar-^{39}Ar ages of Anjar Traps, western Deccan Province (India) and its relation to Cretaceous–Tertiary boundary events. *Current Science* **70**, 990–6.

Wang, D.-N. and Zhao, Y.-N. (1980). Late Cretaceous-early Paleogene sporopollen assemblages of the Jianghan Basin and their stratigraphical significance. *Professional Papers of Stratigraphy and Palaeontology* **9**, 121–71. [in Chinese with English abstract]

Wang, D.-N., Sun, X.-Y., and Zhao, Y.-N. (1990). Late Cretaceous to Tertiary palynofloras in Xinjiang and Qinghai, China. In *Proceedings of the 7th International Palynological Congress*, ed. E. M. Truswell and J. A. K. Owen. *Review of Palaeobotany and Palynology* **65**, 95–104.

Ward, L. F. (1899). The Cretaceous Formation of the Black Hills as indicated by fossil plants. *Nineteenth Annual Report of the US Geological Survey, Part 2*, pp. 521–958.

Watson, J. and Alvin, K. L. (1996). An English Wealden floral list, with comments on possible environmental indicators. *Cretaceous Research* **17**, 5–26.

Wilf, P. (1997). When are leaves good thermometers? A new case for Leaf Margin Analysis. *Paleobiology* **23**, 373–90.

Wilf, P. and Johnson, K. R. (2004). Land plant extinction at the end of the Cretaceous: a quantitative analysis of the North Dakota megafloral record. *Paleobiology* **30**, 347–68.

Wilf, P., Wing, S. L., Greenwood, D. R., and Greenwood, C. L. (1998). Using fossil leaves as paleoprecipitation indicators: an Eocene example. *Geology* **26**, 203–6.

Wilf, P., Johnson, K. R., and Huber, B. T. (2003). Correlated terrestrial and marine evidence for global climate changes before mass extinction at the Cretaceous–Paleogene boundary. *Proceedings of the National Academy of Science* **100**, 599–604.

Wilf, P., Labandeira, C. C., Johnson, K. R., and Ellis, B. (2006). Decoupled plant and insect diversity after the end-Cretaceous extinction. *Science* **313**, 1112–15.

Wilson, G. J., Schiøler, P., Hiller, N., and Jones, C. M. (2005). Age and provenance of Cretaceous marine reptiles from the South Island and Chatham Islands, New Zealand. *New Zealand Journal of Geology & Geophysics* **48**, 377–87.

Wing, S. L. and Boucher, L. D. (1998). Ecological aspects of the Cretaceous flowering plant radiation. *Annual Review of Earth and Planetary Science* **26**, 379–421.

Wing, S. L., Hickey, L. J., and Swisher, C. C. (1993). Implications of an exceptional fossil flora for Late Cretaceous vegetation. *Nature (London)* **363**, 342–4.

Wolbach, W. S., Lewis, R. S., and Anders, E. (1985). Cretaceous extinctions: evidence for wildfires and search for meteoritic material. *Science* **230**, 167–170.

Wolbach, W. S., Gilmore, I., and Anders, E. (1990). Major wildfires at the K–T boundary. In *Global Catastrophes in Earth History; An Interdisciplinary Conference on Impacts, Volcanism, and Mass Mortality*, ed. V. L. Sharpton and P. D. Ward. *Geological Society of America Special Paper* **247**, 391–400.

Wolfe, J. A. (1991). Palaeobotanical evidence for a June 'impact winter' at the Cretaceous/Tertiary boundary. *Nature (London)* **352**, 420–3.

Wolfe, J. A. (1993). A method of obtaining climatic parameters from leaf assemblages. *US Geological Survey Bulletin* **2040**, 1–71.

Wolfe, J. A. and Upchurch, G. R., Jr. (1986). Vegetation, climatic and floral changes at the Cretaceous-Tertiary boundary, *Nature (London)* **324**, 148-52.

Wolfe, J. A. and Upchurch, G. R., Jr. (1987a). Leaf assemblages across the Cretaceous-Tertiary boundary in the Raton Basin, New Mexico and Colorado. *Proceedings of the National Academy of Science* **84**, 5096-100.

Wolfe, J. A. and Upchurch, G. R., Jr. (1987b). North American nonmarine climates and vegetation during the Late Cretaceous. *Palaeogeography, Palaeoclimatology, Palaeoecology* **61**, 33-77.

Yabe, A., Uemura, K., Nishida, H., and Yamada, T. (2006). Geological notes on plant megafossil localities at Cerro Guido, Ultima Esperanza, Magallanes (XII) region, Chile. In *Post-Cretaceous Floristic Changes in southern Patagonia, Chile*, ed. H. Nishida pp. 5-10. Tokyo: Faculty of Science and Engineering, Chuo University.

Young [Yang], C. and Chow [Zhou], M. (1963). Cretaceous and Paleocene vertebrate horizons of north Kwangtung. *Scientia Sinica* **12**, 1411.

Zaklinskaya, E. D. (1977). Pokrytosemennye po palinologicheskim dannym [Angiosperms on the basis of palinological data]. In *Razvitie flor na granitse Mesozoya I Kainozoya [Floral Evolution at the Mesozoic-Cenozoic Boundary]*, ed. V. A. Vakrameev, pp. 66-119. Moscow: Nauka [Geol. Inst. Acad. Nauk SSSR]. [in Russian]

Zhao, Y., Sun, X., Wang, D., and He, Z. (1981). The distribution of Normapolles in northwestern China. *Review of Palaeobotany and Palynology* **35**, 325-36.

Zhao, Z., Ye, J., Li, H., Zhao, Z., and Yan, Z. (1991). Extinction of the dinosaurs across the Cretaceous-Tertiary boundary in Nanxiong Basin, Guangdong Province. *Vertebrata PalAsiatica* **29**, 13-29.

Zhao, Z., Mao, X., Chai, Z., *et al.* (2002). A possible causal relationship between extinction of dinosaurs and K/T iridium enrichment in the Nanxiong Basin, South China: evidence from dinosaur eggshells. *Palaeogeography, Palaeoclimatology, Palaeoecology* **178**, 1-17.

Index